Decision Making Under Pressure

In today's world, the need for effective crisis management extends far beyond traditional emergency response personnel. With an increasing number of people facing high-stakes decisions under pressure, the demand for agile resilient and critical thinking problem-solvers is higher than ever. The aging demographic in emergency services and the high turnover due to burnout makes it challenging to build expertise through experience alone. This book addresses these issues, providing a knowledge base to help individuals and teams operate effectively in volatile, uncertain, complex, and ambiguous environments.

This comprehensive guide covers a range of topics essential for navigating and leading in crises. It offers insights into the trends contributing to today's turbulent landscape and provides strategies for resilient and agile problem-solving. Readers will learn how to lead teams in uncertain conditions, mobilize and manage resources effectively, and enhance coordination among multiple organizations. By using a framework to explore the challenges for the self; the social context and the problem-solving situation, the authors present a series of practical tools to help practitioners navigate through volatility, uncertainty, complexity, and ambiguity to guide action. It will equip any reader with the skill set to confidently tackle incidents, disasters, and crises in whatever setting they find themselves in.

Decision Making Under Pressure: A Crisis Management Sourcebook for Practitioners is an ideal read for practitioners, educators, scholars, and students looking to enhance their decision-making and leadership skills in high-pressure situations, especially those in occupational health and safety, risk management and organizational resilience, ergonomics and human factors, emergency management, crisis leadership, and professional development.

Decision Making Under Pressure

A Crisis Management Sourcebook for Practitioners

Christine Owen and David Parsons

CRC Press
Taylor & Francis Group
Boca Raton London New York

CRC Press is an imprint of the
Taylor & Francis Group, an informa business

Designed cover image: Shutterstock

First edition published 2026
by CRC Press
2385 NW Executive Center Drive, Suite 320, Boca Raton FL 33431

and by CRC Press
4 Park Square, Milton Park, Abingdon, Oxon, OX14 4RN

CRC Press is an imprint of Taylor & Francis Group, LLC

ISBN: 978-1-032-93467-9 (hbk)
ISBN: 978-1-032-93275-0 (pbk)
ISBN: 978-1-003-56600-7 (ebk)

DOI: 10.1201/9781003566007

Typeset in Times
by KnowledgeWorks Global Ltd.

This book is dedicated to our partners, David (CO) and Tamera (DP), with heartfelt gratitude for their support and forbearance. And to Remy, Luca, and Neave – may the world you live in be a better place.

Contents

PART 3 The Situation

List of Figures

List of Tables

List of Tools

Acknowledgments

To the many practitioners we have had the pleasure of working with over the past 10 years across Australia, New Zealand, Canada, and Singapore. You are the inspiration for this book.

About the Authors

Christine Owen's career history spans engagement in academia and industry to help people and organizations engaged in safety-critical work to inquire into, learn from and, if needed, change their work practices. Christine has worked with safety-critical organizations for over 35 years, including the aviation field, emergency medicine, the fire and emergency services industry, and public safety.

Christine researches and facilitates professional development in the areas of human factors and organizational culture; decision making under pressure; debriefing; and leadership and adversity, as well as coaching and mentoring. She has been engaged by industry in several practitioner reviews and evaluations. Christine is passionate about supporting organizations to utilize research outcomes and establish evidence-informed practice. She has over 100 publications in a variety of domains.

David Parsons (ESM) has a passion for creating world-class programs in the organizational resilience areas of risk, crisis decision-making, emergency, and business continuity management. He has served for more than 45 years in a variety of emergency management roles. David is also a Fellow of the Australian Institute of Emergency Services, the Business Continuity Institute, and the Emergency Management Academy of New York, a member of the International Association of Emergency Managers, and an a Life Member of the New South Wales Emergency Service.

David holds a Masters of Emergency Management and degrees in Social Science and Education. David has completed an Emergency Management Fellowship with the Emergency Management Academy in New York, the Leadership in Crises Program with the Harvard Kennedy Business School, and the Lessons Leaned Staff Officer Course with NATO. In 2023, he was awarded an Emergency Services Medal in the Kings Birthday Honours list.

Preface

As researchers, academics, and practitioners working in safety-critical domains for more than three decades, this book emerged from a desire to bring a variety of insights and applications together into one place. Over the past decade, we have shared and taught these ideas across many jurisdictions and countries and in so doing have also learned a lot about the challenges, struggles, and rewards practitioners face when undertaking decision-making in conditions of volatility, uncertainty, complexity, chaos, and ambiguity.

Initially, the participants attending our workshops and Masterclasses came from "traditional" emergency services response agencies: firefighters, police, paramedics had roles as first responders, emergency managers, and commanders. Over the past five years, however, interest has expanded to include many different professions and occupational roles as people recognize the challenges of operating in turbulent conditions. During our conversations, we have wondered what might happen if we took the various programs and content we have developed, designed, and delivered and put them all together in one place. This book has provided the opportunity to do just that.

Preface

As researchers, academics, and practitioners working in safety-critical domains for more than three decades, this book emerged from a desire to bring a variety of models and applications together into one place. Over the past decade, we have shared and taught these ideas across many jurisdictions and countries, and in so doing have also learned a lot about the challenges, struggles, and key factors that influence people when undertaking decision making in conditions of volatility, uncertainty, complexity, chaos, and ambiguity.

Initially, the participants attending our workshops and Masterclasses came from "traditional" emergency services, response agencies, intelligence, police, professions and roles as first responders, emergency managers, and commanders. Over the past five years, however, interest has expanded to include many different professions and occupational roles as people recognize the challenges of operating in turbulent conditions. During our conversations, we have wondered what might happen if we took the various programs and content we have developed, designed, and delivered and put them all together in one place. This book has provided the opportunity to do just that.

Introduction

THE CONTEXT OF MAKING DECISIONS UNDER PRESSURE

If there is anything to be learned from recent and contemporary events, it is that we live in a period of deep uncertainty and that this trend is likely to continue (FERMA, 2025; Finn et al., 2020; World Economic Forum, 2025). In their concept paper exploring the issues of operating within complex socio-technical systems out to 2030 and beyond, Black et al. (2024a) identify several macrotrends that will continue to put high-consequence decision-makers under particular pressure. These include:

- Increasing interconnectivity, multipolarity, and global competition.
- The impact of a changing climate.
- The impact of technological change and digitalization.
- The blurring of domains, both traditional and novel.
- The shifting of international norms and value sets.

They observe that no single or primary trend is driving change or complexity; rather, it is the confluence of multiple factors and their unpredictable interactions that are of greatest concern.

This book provides a knowledge base to support the skills needed to operate effectively in situations of volatility, uncertainty, complexity, and ambiguity (VUCA). While COVID may have played a role in providing a wake-up call for many of us, the need to be agile and resilient problem-solvers continues to escalate (FERMA, 2025). When the issues we face are in a context of turbulence and flux, what is likely to occur cannot be predicted (because of their volatility), and elements are hard to make sense of or understand and thus create uncertainty. In their book *Decision Making Under Deep Uncertainty*, Marchau et al. (2019) characterize deep uncertainty as situations where decision-makers have limited knowledge about the situation and its context and where the knowledge decision-makers need is not available.

Under these circumstances, our previous patterns of problem-solving are unlikely to give us a good way forward. We are not certain how to respond, and there is a lack of historical precedent. Complexity is a situation characterized by many interconnected and moving parts that affect each other. Ambiguity includes unclear or hazy situations. When faced with ambiguity, decision-makers may not even be certain what information to gather, so there is a need for comfort with experimentation. Decision-makers operating in VUCA situations need to have complex sense-making and decision-making skills to sense and respond appropriately to VUCA challenges.

In addition, there is an increasing awareness that the skills needed to cope in contexts of deep uncertainty are not the sole domain of traditional emergency services personnel. More and more of us are called upon to make high-consequence decisions that impact on the well-being of others in time-pressured environments and sometimes under adverse conditions.

A report by FEMA (2024) noted that, while extreme weather events are growing in intensity, frequency, and duration, the challenges faced are not only due to climate change but also changes in human settlements and demography. Heatwaves occur for longer and are more intense. Transitions between drought and flooding are now faster. Pressures on human settlements, health systems, and agriculture are intensified by warming seasons, drought, and migrations of disease and pathogens. Interdependencies of critical lifelines (e.g., reliance on power grids, shallow supply chains, reliance on technology) increase the risk potential for compounding emergencies. Such interconnectivity also exacerbates security challenges in many countries, leading to more diverse, and more complex threats.

Black et al. (2024a) note that when events have elements of complexity, they exhibit emergent properties and often involve non-linearity, hence small changes can have disproportionate or unpredictable effects on the subsequent outcomes. As a result, the environment is often difficult to understand or analyze. Decision-makers are under pressure when they face novel and complex problems where they need to improvise and overcome the limitations of sense-making under these conditions. This means that it will be important for an increasing number of people to navigate uncertain contexts and the challenges they bring to make decisions under pressure.

WHAT THIS BOOK IS ABOUT

In a nutshell, decision-makers and leaders are practitioners who operate in contemporary high-consequence contexts and, we argue, need to think differently in order to successfully interpret and manage high-consequence VUCA situations (Aydiner et al., 2023; Barrett, 2017; Grant, 2023). In the book, readers will find:

- An understanding of the trends contributing to turbulence and deep uncertainty.
- Strategies to support resilient and agile problem-solving.
- Strategies for leading in deep uncertainty and how to mobilize and manage teams in these situations.
- Strategies for stakeholder engagement and building alliances in support of cooperative goals.
- Evidence-based tools that can be applied to address the challenges practitioners face to improve team effectiveness and multi-organizational coordination and to operate with political astuteness to maintain social licence.

Thus, the book is a resource for practitioners, those engaged in the professional development of practitioners, scholars, and students wishing to learn more about

- making decisions under pressure;
- how to lead teams operating in deep uncertainty;
- how to coordinate with affected stakeholders about how to learn from these conditions.

How the Book Is Organized

The book is organized into three parts that highlight the interactive elements that need to be navigated (after Marcus et al., 2015; 2024):

- The self – i.e., the practitioner tasked with making decisions under pressure.
- The social – i.e., the context necessary to work with others.
- The situation – i.e., the challenges associated with a situation suffused with volatility, complexity, ambiguity, and uncertainty.

The Self: Making Decisions Under Pressure

Decision-makers operating under pressure do so frequently without the required information, under time pressure, and in recognition that those decisions have positive and negative consequences for others. In doing so, concerns for the safety and well-being of others bring their own stressors, as does operating in sub-optimal conditions of lack of sleep and fatigue. Making decisions may also be emotionally charged where the decision-maker may be personally impacted by the disaster. In some cases, goals conflict and shift as the event unfolds.

Under certain circumstances, decision-inertia can also become a factor. Waring and Hillon (2025) propose that decision-inertia can be caused by high uncertainty in the context of accountability pressure and the need to select between options with potentially negative outcomes. Anxiety leads to delay or failure in making potentially catastrophic decisions. Demands in this context refer to a variety of aversive physical and psychological stimuli that lead to different emotional states.

In this book, we focus on the stress felt by the individual decision-maker as a practitioner. For those working in high-consequence pressured contexts, stress is almost an inevitable part of the occupation and to survive and thrive at work, decision-makers need to learn about the impacts and how to manage them. Under these circumstances, practitioners need to assemble a nimble response and be ready to adapt to quickly changing conditions. Understanding oneself is thus a critical component of managing effectively.

The Social: Interpersonal Relationships, Teams, and Stakeholders

To add to the challenges, other people are almost always involved. Different interpersonal relationships, levels of expertise, and biases, all affect the quality and timeliness of the management of a high-consequence VUCA-related situation (Owen & Hayes, 2014). Decision-makers and their teams must effectively communicate, coordinate, and integrate their activities and information. Decisions require input from various team members and scattered information, which may vary in quality, quantity, and reliability. Decision-making is often distributed across stakeholders who may have different roles and sometimes competing priorities.

Making decisions under pressure can be complicated because it involves multiple stakeholders, and success relies on their coordination and governance. It is also complicated because there are many causal paths that can lead to potential outcomes. The nature of these paths will partly depend on the specific context.

There are feedback effects between activities and specific outcomes, and some outcomes are emergent and cannot be fully specified or anticipated in advance. Framing decision-making in these contexts as complicated and complex is important for understanding the movement toward desired outcomes and the need to be flexible and adaptable to how these are approached.

The Situation: Volatility, Uncertainty Complexity, and Ambiguity

Szelwach et al. (2023) note that the more volatility, uncertainty, complexity, and ambiguity (or VUCA) in the environment, the harder it is for practitioners to comprehend the situation and articulate a clear way forward. Novel and non-routine problems often have complexity in that they exhibit emergent properties that have been previously outlined. As a result, the environment is often difficult to understand or analyze. VUCA-related conditions require decision-makers to draw on different skills than their expertise from previous patterns might suggest, as well as to build strong stakeholder alliances to develop futures thinking and foresight.

A SOURCEBOOK FOR PRACTITIONERS

The rest of the book is organized around the self/social/situation organizing framework adapted from the work of Marcus et al. (2015, 2024). Following on from the framework of meta-leadership by Marcus et al. (2015) and based on an extensive review of research and the grey literature, we summarize the key attributes practitioners in these contexts require, some of which include those in Table 0.1.

A schematic of the chapter organization is outlined in Figure 0.1.

Chapter Outlines

Chapter 1 sets the situational context and explores the challenges of problem-solving in problem-solving environments suffused with volatility, uncertainty, complexity, and uncertainty. It considers complex and wicked problems and the leadership approaches practitioners need to navigate, which include embracing a collaborative learning mindset. It provides a tool that assists the practitioner in identifying elements contributing to volatility in the problem-solving situation.

Part 1: The Self

Chapters 2–7 pick up the issues for the individual tasked with making decisions under pressure to better understand the challenges of working in VUCA conditions. Chapter 2 outlines the kinds of thinking styles individuals typically employ which include rational (deliberative and analytical) and intuitive (sub-conscious, using decision heuristics). The chapter highlights some of the pitfalls and introduces some strategies to enhance decision-making capabilities, including metacognition, decision control processes, and situational awareness techniques to effectively navigate VUCA environments. In addition to these strategies, the chapter concludes with a tool to self-assess sources of pressure at work.

TABLE 0.1
Attributes for Practitioners Making Decisions Under Pressure

Self	Social	Situation
• Humble and comfortable with challenge	• Building diverse and effective teams including human and nonhuman agents	• Analyses and manages uncertainty and disorder
• Open-minded and curious	• Engages in a range of perspectives to stress-test emergent thinking	• Dynamic planning in deep uncertainty
• Self-aware and self-disciplined	• Acting as an emotional shield to protect the team	• Monitoring proactively for early warning signals
• Meta-cognitive awareness of how one's own frames influence perspective-taking	• Diagnoses team dysfunction	• Uses technology to support sense-making and debate
• Emotionally intelligent and politically astute	• Strong stakeholder engagement	• Uses foresight and feedback loops for planning adjustments
• Acts with integrity and ethics	• Dynamic delegation and mentoring	• Extemporizes protocols and suspends rules where necessary
• Active in listening	• Organizationally intelligent	• Constructively works with differences and conflicts of interest
• Comfort in conflict	• Negotiation and conflict resolution to expand disparate constituencies	
• Ability and willingness to change decisions	• Actively seeking out alliances and partnerships	
• Self-compassion		
• Empathically attuned to others' emotional states		

Source: Alshayhan et al. (2024), Black et al. (2024b), Boyatzis et al. (2017), Collins et al. (2023), Eburn and Gissing (2024), Feldmann-Jensen et al. (2019), FEMA (2024), Klein et al. (2006), Mahmud et al. (2020), Marcus et al. (2015), Wilson (2023), Yates and Hartley (2021), see also references in individual chapters.

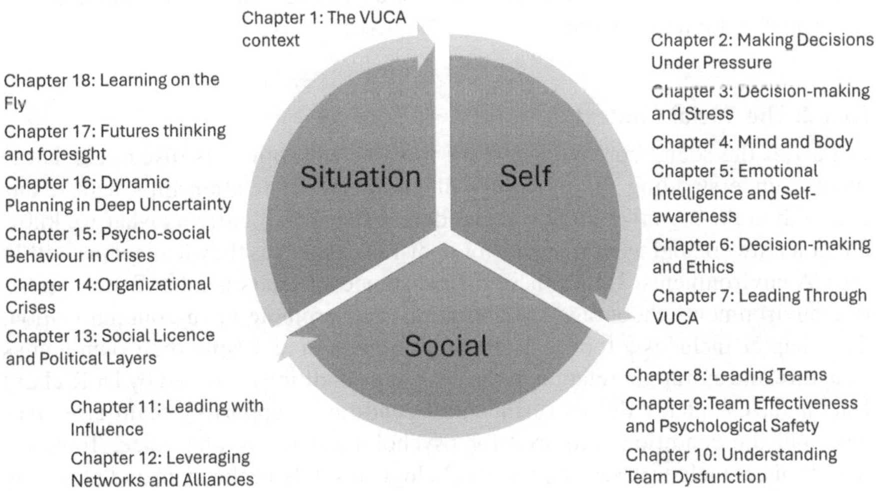

FIGURE 0.1 Outline of chapters in the book

In Chapter 3, the impact of stress on decision-making is explored. The chapter discusses the deleterious effects of an amygdala hijack on decision-making and offers strategies and tools to support stress management. Chapter 4 discusses how decision-making is informed not through rational thinking but through the information that is shared between the gut and the brain along the vagus nerve. The chapter outlines the research linking gut microbiota to mood and cognition and identifies strategies to improve gut health. It concludes with a discussion about the relationship between the body and the mind and the importance of both rest and sleep to support resilience and effective decision-making.

Chapter 5 takes up the importance of emotional intelligence and self-awareness for the individual decision-maker. It highlights Daniel Goleman's four domains of emotional intelligence: Self-Awareness, Self-Management, Social Awareness, and Relationship Management, and links this to research with wildland firefighters and what they look for in their leaders. Emotional triggers that can derail decision-making effectiveness are also discussed. The chapter outlines some strategies for practitioners, which include becoming mindful about the differences between being in a protective mode and reframing emotional triggers for adaptation. The chapter concludes with a tool to support practicing self-compassion.

Chapters 6 and 7 explore the implications for leadership in terms of both ethical decision-making (Chapter 6) and implications for the individual who takes on a leadership role under VUCA conditions. In Chapter 6, a tool is provided to guide decision-makers in ethical considerations. Strategies to support ethical decision-making such as mindful listening are outlined. The chapter also outlines some of the ethical considerations in the application of artificial intelligence into decision-making in high-consequence VUCA-related contexts. Chapter 7 takes up some of the challenges that can impact the decision-maker in their individual approach to leadership. A tool is offered to support increased awareness about some of the traps leaders can fall into. It also provides some strategies to support articulating uncertainty and legitimizing doubt.

Part 2: The Social Context

In Part 2, the social context necessary to work with others is discussed. These involve understanding the dynamics of interpersonal relationships with teams and with multi-organizational stakeholders. Chapter 8 highlights what we know about teamwork and what it means for decision-makers as they lead teams within VUCA environments. Under these circumstances, leaders need to foster a positive environment, encourage collaboration, and promote open communication. The chapter includes a tool to assist practitioners in assessing their own team's structure and culture in relation to the enabling conditions outlined by Dr Richard Hackman. Chapter 9 delves further into conditions supporting team effectiveness and the conditions required for psychological safety and voice. Tools for practitioners include assessing the psychological safety of the team and assessing the quality of team-talk in support of psychological safety. It also discusses some of the challenges in communication in hierarchical teams and outlines the PACE protocol, used in healthcare (and applied in emergency management), to escalate concerns and promote collaborative problem-solving as a means of lowering the power-authority gradient.

Chapter 10 continues with an exploration of challenges in team communication by exploring the reasons why teams are sometimes dysfunctional. The five dysfunctions of a team promoted by Lencioni (2006) are discussed together with strategies decision-makers and leaders can use to overcome them. A tool practitioners can use to assess their own leadership communication styles within the team is also provided.

Chapter 11 moves beyond the team to consider the importance of strategic communication and stakeholder engagement in decision-making and leadership. The chapter discusses strategies to resolve conflict and foster collaboration among diverse stakeholders. A tool to assist practitioners in the art of crafting narratives and storytelling to support influence is outlined. Chapter 12 continues to consider strategies for stakeholder engagement and discusses strategies to leverage networks and alliances. A tool to analyze stakeholders in terms of their potential power and interest is provided to support the prioritization of attention to stakeholders.

Part 3: The Situation

In Part 3, the situation, approaches, and strategies that can be used in problem-solving environments suffused with volatility, complexity, ambiguity, and uncertainty are explored.

Chapter 13 discusses situations where the crisis is at an organizational level and considers the implications for leadership. It frames organizational crises as originating either internal or external to the organization and whether it was based on an intended or unintended act.

Chapter 14 considers what it means for practitioners operating under conditions of deep uncertainty in how they navigate such contexts. This includes analyzing the nature of uncertainty and engaging in horizon-scanning. Strategies for practitioners to use when managing risk are outlined together with a tool to assess trigger points and the courses of action required. Chapter 15 continues the discussion around futures thinking and scenario planning including ways to enhance creative thinking. Strategies for practitioners to apply foresight are provided. A tool to conduct a pre-mortem is outlined.

Chapter 16 considers the impact of VUCA-related conditions on the psycho-social human behavior of the public. Understanding the psycho-social reactions of affected communities is important for practitioners. A tool is provided for practitioners to assess the potential for psycho-social trauma within impacted communities. Chapter 17 explores the strategies needed by practitioners when they navigate between the worlds of bureaucratic order (e.g., applying legislation, protocols, and standards) and the chaos that can be part and parcel of facilitating locally led decision-making in impacted communities. Finally, Chapter 18 takes up the issue of social license and political layers engaged in VUCA-related environments. Strategies for practitioners to develop political astuteness are included as well as a tool to analyze the political nature of a VUCA event.

The book's conclusion discusses the implications for the future and the ways in which each of the chapters included in the book provides some insights and skills likely to be needed.

MAIN POINTS

- There is an increasing amount of turbulence and flux in many of our workplaces which means that operating in conditions of volatility, uncertainty, complexity, and ambiguity (VUCA) is now a new normal for many of us, not just those working in traditional emergency services organizations.
- Interdependencies and tightly coupled systems are leading to the impacts in one domain being felt on a wider scale.
- The book is intended for practitioners operating in VUCA-related conditions. It is organized into three parts. These include
  - The self, i.e., the individual tasked with making decisions under pressure
  - The social, i.e., the context necessary to work with others, and
  - The situation, i.e., the challenges associated with a context suffused with volatility, complexity, ambiguity, and uncertainty.
- The contents of various chapters comprising the book are outlined.

REFERENCES

Alshayhan, N., Hill, S., Saitgalina, M., & Yusuf, J. E. (2024). Leadership in collaborative emergency management for compound hurricane-pandemic threats: Insights from practitioners' experiences. *Risk, Hazards & Crisis in Public Policy, 15*(3), 328–346.

Aydiner, C., Corbin, T. B., & Tan, C. (2023). Developing critical thinking and effective communication skills in the future aviation workforce. *National Training Aircraft Symposium (NTAS). 37.* Retrieved June 2, 2025, from https://commons.erau.edu/ntas/2022/presentation/37

Barrett, L. F. (2017). *How emotions are made: The secret life of the brain.* Pan Macmillan.

Black, J., Lucas, R., Kennedy, J., Hughes, M., & Fine, H. (2024a). *Command and control in the future: Concept paper 1: Grappling with complexity.* Rand Corporation. rand.org/pubs/research_reports/RRA2476-1.htm

Black, J., Lucas, R., Kennedy, J., Hughes, M., & Fine, H. (2024b). *Command and control in the future. Concept Paper 1: Grappling with complexity.* The Global Strategic Partnership. RAND Europe. Retrieved June 2, 2025, from https://www.rand.org/content/dam/rand/pubs/research_reports/RRA2400/RRA2476-1/RAND_RRA2476-1.pdf

Boyatzis, R. E., Thiel, K., Rochford, K., & Black, A. (2017). Emotional and social intelligence competencies of incident team commanders fighting wildfires. *The Journal of Applied Behavioral Science, 53*(4), 498–516.

Collins, M. D., Dasborough, M. T., Gregg, H. R., Xu, C., Deen, C. M., He, Y., & Restubog, S. L. D. (2023). Traversing the storm: An interdisciplinary review of crisis leadership. *The Leadership Quarterly, 34*(1), 101661.

Eburn, M., & Gissing, A. (2024, November). *Principles of best practice strategic crisis management arrangements for catastrophic disasters.* Technical Report, Natural Hazards Research Australia. Retrieved June 2, 2025, from https://naturalhazards.com.au/sites/default/files/2024-11/Crisis%20Management%20Arrangements%20Final%20Report%20November%202024.pdf

Feldmann-Jensen, S., Jensen, S. J., Smith, S. M., & Vigneaux, G. (2019). The next generation core competencies for emergency management. *Journal of Emergency Management, 17*(1), 17–25. https://doi.org/10.5055/jem.2019.0393

FEMA. (2024, November). *Strategic foresight 2050, final report.* Retrieved June 2, 2025, from https://www.fema.gov/sites/default/files/documents/fema_strategic-foresight-2050-final-report.pdf

FERMA (2025). Federation of European Risk Management Associations, Anticipating Changes Shaping the future, Brussels. https://www.ferma.eu/wp-content/uploads/2025/02/20250217-FERMA-ECIIA-webinar-Global-Risks-Report-2025.pdf

Finn, P., Mysore, M., & Usher, U. (2020). *When nothing is normal: Managing in extreme uncertainty*. Retrieved September 13, 2022, from https://www.mckinsey.com/business-functions/risk-andresilience/our-insights/when-nothing-is-normal-managing-in-extreme-uncertainty

Grant, A. (2023). *Think again: The power of knowing what you don't know*. Penguin.

Klein, K. J., Ziegert, J. C., Knight, A. P., & Xiao, Y. (2006). Dynamic delegation: Shared, hierarchical and deindividualized leadership in extreme action teams. *Administrative Science Quarterly*, *51*, 590–621. http://dx.doi.org/10.2189/asqu.51.4.590

Lencioni, P. (2006). *The five dysfunctions of a team*. John Wiley & Sons.

Mahmud, A., Mohammad, Z., & Abdullah, K. A. (2020). Leadership in disaster management: Theory versus reality. *Journal of Clinical and Health Sciences*, *5*(1), 4–11.

Marchau, V. A., Walker, W. E., Bloemen, P. J., & Popper, S. W. (2019). *Decision making under deep uncertainty: From theory to practice*. Springer Nature. https://doi.org/10.1007/978-3-030-05252-2_1

Marcus, L. J., Dorn, B. C., Henderson, J., & McNulty, E. J. (2015). *Meta-leadership: A framework for building leadership effectiveness*. National Preparedness Leadership Initiative. https://cdn1.sph.harvard.edu/wpcontent/uploads/sites/2443/2020/01/Meta-leadership-Overview-Working-PaperFinal.pdf

Marcus, L. J., McNulty, E. J., & Grimes, J. O. (2024). Crisis meta-leadership and the practice of disaster medicine. In G. Ciottone (Ed.) *Ciottone's disaster medicine* (3rd ed., pp. 269–274). Elsevier.

Owen, C., & Hayes, P. (2017). Human factors in emergency management. In *Human factors challenges in emergency management* (pp. 1–18). CRC Press.

Szelwach, C., Sweet, K. M., Vermeer, P. E., & Tarhini, K. M. (2023). A holistic approach to embodied leadership development at the US Coast Guard Academy. *Organization Development Journal*, *41*(1).

Waring, S., & Hillon, B. (2025). Examining what factors affect firefighter decisions to declare a major incident. *International Journal of Disaster Risk Reduction*, *116*, 105098.

Wilson, S. (2023). Leadership in a VUCA context: Some foundational considerations. *Journal of Applied Journalism & Media Studies, Special Issue 'Communicating Chaos: New Perspectives on VUCA*, *12*(2), 169–183. https://doi.org/10.1386/ajms_00112_1

World Economic Forum (2025). Global Risks Report, from Global Risks Report 2025 | World Economic Forum.

Yates, S., & Hartley, J. (2021). Learning to lead with political astuteness. *International Public Management Journal*, *24*(4), 562–583. https://doi.org/10.1080/10967494.2021.1894281

1 The VUCA Context

INTRODUCTION

All decision-making and leadership action takes place in a specific context. While the acronym volatility, uncertainty, complexity, and ambiguity (VUCA) is not new, it speaks to the existing contexts that many practitioners and leaders face in contemporary times.. Sometimes there is an additional C included (i.e., VUCCA) which stands for chaos. Part of the reason why decision-makers are under pressure is because sense-making and subsequent actions are embedded within these kinds of contexts and interact with the dynamics of the situation.

When a situation is volatile, it implies that changes are rapid with likely significant impacts and for which effective and easy-to-adopt resolutions or mitigations are not readily available (Wilson, 2023). In these cases, it is easy for the disruptive effects of decisions to ripple out with sometimes unintended consequences, including cascading and compounding problems.

When the decision-making context is uncertain, then it is very difficult to predict with any real confidence what the outcome of decisions taken might be. Under these circumstances, the pathway to resolution is not clear or simple. The context is likely to be exacerbated by inaccurate information, information overload, or intentional disinformation.

These conditions make problem-solving more complex. In our current world, many systems are interconnected, involving different stakeholders and communities. What the problem is and how it should be resolved will differ depending on which stakeholder perspective is considered. This means there is likely to be high levels of contestation in how the problem or situation is perceived and what various stakeholders think ought to be done (Wilson, 2023). This means that in VUCA environments, the problems faced can be both complex and wicked, as discussed later in this chapter.

The ambiguity present in the decision-making context means that the situation and its interpretation remain hazy. Making sense of what is happening and what needs to be done is not straightforward, even though some stakeholders may think that's the case. In a nutshell, facts and assumptions cannot be taken for granted (Wilson, 2023).

Decision-making in VUCA-related contexts will be ones where disruption, inconsistencies, and tensions must be navigated. As the literature repeatedly shows, under these conditions, attempting to solve problems with command-and-control decision-making structures frequently fails (O'Rourke & Leonard, 2018; Owen et al., 2015, 2017).

As later chapters will discuss, how we make decisions is influenced by a range of physiological and psychological factors. When we are under pressure, there is a tendency to revert to previous solutions which may not be appropriate in the context.

DOI: 10.1201/9781003566007-1

This is because, in framing the problem to be solved, we try to pattern-match it to something we've seen before. What is needed in VUCA-related contexts is often creativity and innovation, which will be challenging when facing problems that are not simple and instead are novel and complex.

SIMPLE AND COMPLEX PROBLEMS

When a problem is simple, there is typically clarity around the cause and the outcome, and a known solution can be applied. As the problem-solving context grows in complexity, the number of likely stakeholders affected increases along with their diversity of perspectives. This requires a different leadership approach to decision-making. In a simple, critical situation, for example, decisions made using a command-and-control problem-solving approach will be appropriate. However, as the complexity and diversity of stakeholders and perspectives increase, a more political, negotiated approach is required.

This is illustrated in Table 1.1. On the one hand, there is the degree of complexity involved: that is: Is the problem one that is known and for what there are already solutions in place that can be trained for and applied? Or is the problem novel and where the solution is not known? There are likely to be differing perspectives on both the source of the problem and its proposed solution. Tame problems are those that only impact a single party, community, or organization. Wicked problems are those that impact multiple parties where various stakeholders are likely to have differing interpretations of the problem and preferred solutions.

Wicked Problems

The nature of VUCA-related situations is that practitioners are frequently faced with solving problems that are wicked because they are both novel and have impacts on many stakeholders. This makes them both complicated and complex. According to

TABLE 1.1
Effect of the Complexity of Problems and Diversity of Stakeholders

Complexity	Diversity		
	Single Party	Multiple Parties Each Having Only Some Knowledge	Multiple Parties Conflicting in Values and Interests
Neither problem nor solution known	–	Wicked	Very wicked
Problem known solution not known	Almost tame	–	Wicked
Problem and solution known	Tame	Almost tame	–

Source: Alford and Head (2017).

the Australian Public Service Commission (APSC, 2007), wicked problems share the following features.

Wicked problems can be difficult to clearly define. The nature and extent of the problem depend on who has been asked to explain the nature of the problem. That is, different stakeholders have different versions of what the problem is.

Wicked problems have many interdependencies and are often multi-causal. Also, there are often internally conflicting goals or objectives within the broader wicked problem.

Attempts to address wicked problems often lead to unforeseen consequences. Because wicked problems are multi-causal with many interconnections to other issues, it is often the case that measures introduced to address the problem lead to unforeseen consequences elsewhere.

Wicked problems are often not stable. Frequently, a wicked problem and the constraints or evidence involved in understanding the problem (e.g., legislation, scientific evidence, resources, political alliances) are evolving while decision-makers are trying to address the problem.

Wicked problems usually have no clear solution. Since there is no definitive, stable problem, there is often no definitive solution to wicked problems.

Wicked problems are complex. It is often the subtle nuance of wicked problems, rather than their technical complexity that is challenging. Complex disasters, like wicked problems, hardly ever sit conveniently within the responsibility of any one organization. They often require action and cooperation of multiple parties, which will require a coordinated approach (see Chapter 11).

A Collaborative Learning Mindset

To achieve advances in wicked problems, the decision-maker will need collaboration and a learning mindset (Parker & Barlett, 2008). The leadership approach is not so much of a traveler who knows the route, but more of an explorer who has a sense of direction but no clear route. Search and exploration, watching out for possibilities and inter-relationships, however unlikely they may seem, are part of the practice of leadership under these circumstances. There are ideas as to the way ahead, but some may prove abortive. What is required is a readiness to see and accept this, rather than to proceed regardless on a path that is found to be leading nowhere or in the wrong direction (APSC, 2007).

STRATEGIES FOR PRACTITIONERS

Strategies to navigate VUCA-related environments comprising wicked problems include:

- Ensuring collaboration across boundaries, including engaging community members.
- Stimulating discussion about the different options available.
- Encouraging thinking about innovative ways to address the problem.

- Fostering a culture that encourages discussion about contentious issues among stakeholders.
- Identifying a principles-based approach, rather than sticking with specific policies and rules.

Framing VUCA Events

While this book examines leading and managing disaster events, the ideas here are also relevant to a range of other contexts, including crises impacting organizations (i.e., the organizations' survival) and/or the well-being of communities (e.g., malicious attacks such as cyber-attacks on critical infrastructure) or operating in situations with intentional disinformation about an emerging problem. It is helpful then to consider what is meant by different terms that often characterize operating in such conditions. The difference between what might be termed, for example, an emergency, disaster, crisis, or indeed catastrophe is one of novelty and scale (Quarantelli, 2006).

Incidents, Disasters, Crises, and Catastrophes

Several bodies (e.g., FEMA, National Emergency Management Australia) have considered the differences between emergency events, disasters, and catastrophes. There are several organizations that deal with emergency events as business-as-usual (e.g., fire services). In this respect, those events are sometimes called incidents or emergency events. An emergency event can be routine or non-routine. Routine events have pre-determined plans in which responders and managers have been trained, and in many cases, the operation is successful. While decisions are made that may have significant consequences, there is unlikely to be high pressure due to the training and standard operating procedures involved.

In contrast, a disaster is one where local organizations are overwhelmed and in need of support. Given interdependencies between systems, impacts in one system can trigger failures in another leading to cascading disaster consequences (Comfort, 2005). Disasters are a subset of crises (Marcus et al., 2024). Crises are "high-impact events that present a serious threat to the basic structures or fundamental norms of a system" (Marcus et al., 2024, p. 270). As such, they have the potential to disrupt everyday life. They also have the potential to destroy the reputation of organizations and the individuals who lead or work within them. Within communities, crises can lead to fear and a loss of confidence that those making decisions know what they are doing.

A key feature of a catastrophe is its novelty. The challenges this presents are what distinguish them from everyday incidents or emergency events (O'Rourke & Leonard, 2018). Under these circumstances, pre-determined plans often fail because of the complexity involved. Sometimes decisions taken have interactive effects with the event where a decision leads to an unanticipated outcome. To understand and manage making decisions under pressure, we need to recognize the challenges of working in such contexts as well as the limitations for individuals, organizations, and communities.

Naming Up Volatility in Problem-Solving Environments

One of the challenges in making decisions under pressure is recognizing the amount of volatility, uncertainty, complexity, and ambiguity present in an environment. As most of our organizational processes that guide our activities are based on what is already known and has been previously experienced (e.g., routines in standard operating procedures), we need a language to articulate VUCA conditions to assist ourselves and others to recognize that the situation is different from ones we might have faced in the past.

Tool 1.1 has been adapted from the work of Dirk Maclean and his book *Shoot, Don't Shoot: Minimising the Risk of Catastrophic Error through High Consequence Decision Making* (2017). Maclean's work provides us with an assessment tool that can examine the different components of high-consequence decision-making. Indicators have been extracted and can give an indication of the kinds of challenges involved in a high-consequence decision-making situation by breaking down the various factors that will influence the complexity for the practitioner.

We have adapted his ideas here to develop a tool that we believe helps to provide indicators of the amount of volatility, uncertainty, complexity, and ambiguity in a situation. The following statements included in Tool 1.1 can help in identifying the level of VUCA involved in the event and its sources. These have implications for the degree of pressure due to the high consequences and stakes involved. In using this tool the more ticks, the more VUCA is present in a situation.

Consider an event you faced or are currently facing and conduct an assessment of the various aspects indicated below.

TOOL 1.1 IDENTIFYING VOLATILITY IN THE DECISION-MAKING CONTEXT

i. Information management

Factor	Tick if present	Comment
Information is not available, gaps exist	○	
Information volume is overwhelming	○	
Information is out of date, being overtaken by events	○	
Information is unreliable	○	
Information is conflicting	○	
Information is ambiguous	○	

ii. Situation assessment

Factor	Tick if present	Comment
Situation is dynamic, complex, non-linear	○	
Situation is out of control, minimal ability to influence events	○	
Situation is unique, without precedent	○	
Situation is unassessed up to now, and unexpected	○	
Situation is unfamiliar, lack of personal experience	○	
Situation appears chaotic, patterns are hard to find	○	
The significance of individual pieces of information is hard to evaluate	○	
Multiple assessments are possible and credible	○	
Time for assessment is limited and excludes structured techniques	○	

iii. Role of actors

Factor	Tick if present	Comment
Ad hoc community response groups forming	○	
Community members not heeding advice of emergency services	○	
Significant media presence and scrutiny	○	

iv. Mission and objectives

Factor	Tick if present	Comment
Current objectives are vague and/or irrelevant	○	
Leader has not stated their intent	○	
Leader's intent does not address the actual decision that needs to be made	○	
Desired end state is not clear	○	
Desired end state is not achievable, least-worst options need to be considered	○	
Operational necessity clashes with the need for political or community support	○	

v. Constraints

Factor	Tick if present	Comment
Guidance on acceptable risk is not available	○	
Own capabilities (multi-agency) not fully understood, team or equipment is new	○	
Availability of assistance is unknown	○	

vi. Decision-making authority

Factor	Tick if present	Comment
Unclear whether authority to make a decision has been granted	○	
Some options are authorized; others require permission	○	
Contact with the chain of command is broken	○	

vii. Anticipation and preparation

Factor	Tick if present	Comment
Situation was not foreseen	○	
Warning signs were not picked up	○	
Possible courses of action were not discussed in advance	○	
Decision support, guidance, and procedures are not available	○	

viii. Consequences

Factor	Tick if present	Comment
Consequences are hard to foresee	○	
Reactions of the media/community are hard to predict	○	
Unintended consequences are likely to occur	○	
Consequences are difficult or impossible to reverse	○	

ix. Urgency

Factor	Tick if present	Comment
Immediate decision is needed	○	
One single decision will have a large influence on the outcome	○	

x. Stakes

Factor	Tick if present	Comment
There is potential for catastrophic error	○	
There is a high likelihood of personal threat	○	
There is a high likelihood of a threat to others	○	

xi. Other factors degrading decision-making ability

Factor	Tick if present	Comment
Stress levels are high	○	
Decision-makers are fatigued	○	
Decision-makers are distracted	○	
Decision-makers are suffering cognitive overload	○	

Source: Maclean (2017).

The tool provides a subjective assessment of indicators of volatility in a problem-solving situation. The indicators may also be useful in articulating the dynamics of the situation to others to assist them in accepting that the approaches taken may need to be reassessed in light of the volatility present.

MAIN POINTS

The difference between incidents, disasters, crises, and catastrophes is one of novelty, scale, and how it is managed. Effective decision-making in VUCA environments requires a paradigm shift from traditional command-and-control approaches. The complexities of VUCA contexts, often characterized by wicked problems, demand a collaborative learning mindset. This involves embracing exploration, adaptability, and cooperation among diverse stakeholders. Key strategies for practitioners include:

- **Shift from command-and-control:** Adopt collaborative, innovative approaches to navigate VUCA environments.
- **Recognize VUCA factors:** Utilize assessment tools to identify the degree of VUCA present in a situation.
- **Prioritize collaboration and nuance:** Evolve solutions in response to emerging complexities, considering multiple stakeholder perspectives.
- **Foster a learning mindset:** Encourage exploration, adaptability, and cooperation to effectively address novel, multi-causal problems with unforeseen consequences.

REFERENCES

Alford, J., & Head, B. W. (2017). Wicked and less wicked problems: A typology and a contingency framework. *Policy and Society*, *36*(3), 397–413. https://doi.org/10.1080/14494035.2017.1361634

Australian Public Service Commission (2007). Tackling wicked problems: A public policy perspective. Retrieved April 26, 2005, from https://www.enablingchange.com.au/wickedproblems.pdf

Comfort, L. K. (2005). Risk, security, and disaster management. *The Annual Review of Political Science*, *8*(1), 335–356.

Maclean, D. (2017). Shoot, don't shoot: Minimising risk of catastrophic error through high consequence decision making. Air Power Development Centre

Marcus, L. J., McNulty, E. J., & Grimes, J. O. (2024). Crisis meta-leadership and the practice of disaster medicine. In G. Ciottone (Ed.), *Ciottone's disaster medicine* (3rd ed., pp. 269–274). Elsevier.

O'Rourke, S., & Leonard, H. B. (2018). Operating at the speed of challenge: Adaptive C2 as a new paradigm or police command and control at critical incidents. PCL Discussion Paper, Harvard Kennedy School Program on Crisis Leadership. Retrieved April 26, 2025, from https://www.hks.harvard.edu/sites/default/files/centers/research-initiatives/crisisleadership/files/ORourke_Leonard_OperatingAtTheSpeedOfChallenge_web20180611.pdf

Owen, C., Scott, C., Adams, R., & Parsons, D. (2015). Leadership in crisis: Developing beyond command and control. *Australian Journal of Emergency Management*, *30*(3), 15–19. https://knowledge.aidr.org.au/media/1472/ajem-30-03-05.pdf

Owen, C., Scott, C., Adams, R., & Parsons, D. (2017). *Beyond command and control: Leadership, culture and risk.* Taylor and Francis.

Parker, S., & Barlett, J. (2008). *Towards agile government.* State Services Authority (Vic). Victoria, Australia. Retrieved April 26, 2025, from https://apo.org.au/sites/default/files/resource-files/2008-05/apo-nid3888.pdf

Quarantelli, E. L. (2006). *The disasters of the 21st century: A mixture of new, old, and mixed types.* Preliminary Paper #353. Disaster Research Center, University of Delaware. Newark, DE.

Wilson, S. (2023). Leadership in a VUCA context: Some foundational considerations. *Journal of Applied Journalism & Media Studies, Special Issue 'Communicating Chaos: New Perspectives on VUCA'*, *12*(2), 169–183. https://doi.org/10.1386/ajms_00112_1

Part 1

The Self

To successfully negotiate VUCA events, practitioners need a strong understanding of their strengths as well as the potential risks and error traps they may fall into simply because they are human.

Chapter 2 explores how practitioners engage in both rational and intuitive forms of thinking and how cognitive biases can sometimes operate as blind spots. Under these circumstances, it is important to engage in both metacognition and meta-recognition to monitor decision-making in challenging situations. Chapter 3 then takes up the role of stress and its impact on decision-making. Self-awareness is important here and can be used as a resource to identify when resilient strategies need to be employed in order to maintain effectiveness.

Chapter 4 picks up on an increasingly important area for research that pertains to the gut microbiome. It contends that the body is just as important as the brain in monitoring signals (such as if external stimuli were expected or a threat) and is equally important for self-care when operating under duress. For example, engaging in strategies to support rest and sleep is key in effective decision-making.

Chapter 5 looks at the relationship between emotional intelligence and self-awareness, making the case for why understanding emotional self-awareness is necessary for effective decision-making. This includes practitioners being mindful of compassion fatigue and the importance, when under pressure, of actively practicing self-compassion.

Chapter 6 considers aspects that are important for the practice of ethical decision-making. It outlines strategies for practitioners to embed ethical principles into decision-making.

DOI: 10.1201/9781003566007-2

Chapter 7 identifies some of the challenges for self-as-leader when operating in deep uncertainty. It invites the practitioner to identify strategies to overcome common mind traps.

Thus, the chapters in this part are focused on assisting the practitioner to better understand their own ways of operating, the triggers to be aware of, and some of the strategies to engage in when making decisions under pressure.

2 Making Decisions Under Pressure

INTRODUCTION

Decision-making is the process of evaluating alternatives and selecting a course of action. When we make decisions, we are typically making choices between several options available, each of which might lead to different outcomes, and in turn lead to different consequences (Sarmiento et al., 2024). Decision-makers typically choose between different courses of action based on the assessment of the consequences (Starcke & Brand, 2012). As discussed in the introduction and Chapter 1, the contexts facing many decision-makers are increasingly volatile, uncertain, complex, and ambiguous. Under these conditions, it is often the case that decision-makers are forced to make decisions without all the needed information under time-constrained pressure and that those decisions may lead to sub-optimal outcomes for others. Paul 't Hart summed it up nicely in the following quote. Decision-makers…

> need to take highly consequential decisions in a context in which they do not have all the numbers, they can't delegate the issues to a commission and can't get the experts to study it for a few months. They have to act much faster than governments normally act. And often that acting involves doing quite unpleasant things, or disappointing a lot of people, or making tough decisions about the allocation of scarce resources.
>
> *('t Hart, 2014, p. 172)*

When decisions are made in high-consequence situations, the pressure and subsequent stress come from the volatility and uncertainty of those conditions. When conditions are novel and non-routine with outcomes with potentially high stakes, decision-makers are faced with the stress of deep uncertainty. While the impact of stress on cognition (thinking) will be discussed in more detail in the following chapter *Decision-making and Stress*, what is important to note here is how stimuli in decision-making are influenced by the flood of hormones and neurotransmitters that arise when confronted with volatility and uncertain factors. In short, stress is a reaction to fear, and fear is a reaction to uncertainty.

Decision-making is mediated by the degree to which the amygdala is activated. The amygdala, a small set of nuclei at the base of the brain, modulates the fear response. While the jury is still out about whether the activation of the amygdala increases or reduces aversion to losses (Molins et al., 2021), what is well established is that decision-making under pressure is likely to be more reactive and focused on immediate gains at the expense of greater consideration of longer-term consequences. Therefore, it is important to understand how we think and how it can sometimes catch us out.

DOI: 10.1201/9781003566007-3

THINKING ABOUT THINKING

Typically, the way individuals make decisions has been characterized as either rational or intuitive. A rational decision-making style is described as employing the following steps: (i) defining the problem; (ii) diagnosing the problem and its potential solution; (iii) weighing up several alternative courses of action; (iv) choosing the best option; and (v) enacting the selected option. An intuitive thinking style is the one we use when we come to a decision without conscious calculation (called decision-heuristics by Tversky and Kahneman in 1974).

Gary Klein (1997a, 1997b, 2011) has championed this sub-conscious thinking style as "recognition-primed decision-making" (or RPD). It is the kind of thinking experts engage in when they are sizing up a situation. Rather than consciously putting all the pieces of the puzzle together, they are holistically reading a situation to compare it with a mental slide carousel of images based on previous experiences to categorize the situation and consider appropriate courses of action. The difference in these thinking styles has popularly been called thinking fast and thinking slow (see Kahneman, 2011; Tversky & Kahneman, 1974). That is, intuitive sub-conscious thinking as "system 1" (thinking fast) and thinking in a conscious, deliberative and analytical way is often referred to as "system 2" thinking (thinking slow).

In upcoming sections of this chapter, we will delve into various tools and strategies designed to enhance your decision-making capabilities while highlighting common pitfalls to avoid. This awareness will enable you to make the most effective decisions within the constraints of time and context.

A key insight from recent literature on threat assessment and decision-making (Penney et al., 2022) indicates that many decisions in volatile, uncertain, complex, and ambiguous (VUCA) (see Chapter 1) environments tend to be naturalistic, meaning they are often made quickly and based on intuition rather than a formalized process. This can be both an advantage and a disadvantage, as intuitive decisions can leverage experience and insights but may also bypass thorough analysis.

The review by Penney et al. (2022) emphasizes that step-based protocols designed to assist decision-making are most effective when there is an already established organizational solution to a problem. That is, rational step-based protocols work well for problems that are known and less well when problems are novel and complex. As we explore these concepts further, we'll look at specific tools and frameworks that can help you develop a structured approach to decision-making, ensuring you're prepared to tackle complexities in VUCA environments effectively.

Studies demonstrate that experts and novices use different strategies for establishing situational awareness (Penney et al., 2022). Novices are likely to be explicitly processing information in a deliberative way and using rule-based procedures (system 2 thinking) to work through problem-solving, whereas experts will be using unconscious and intuitive processes (system 1 thinking), drawing information from multiple sources including experience. In this way, experts are sometimes referred to as thinking in decision-making cycles. That is, according to Orasanu and Connolly (1993), experts "think a little, act a little, and then evaluate outcomes and act some more" (p. 19).

Research conducted by Butler et al. (2023) and Cohen-Hatton et al. (2015), with emergency services decision-makers in the UK, has found that instead of working through a prescribed model of decision-making, that is, to first assess the situation, formulate a plan, and execute on the plan, the decision-makers frequently assessed a situation then executed an action and missed formulating a plan.

Moreover, they observed that they frequently made decisions in the "here and now" without anticipating the development of the event. In training, Cohen-Hatton et al. (2015) developed what they called a "decision control process". A decision control process is a rapid mental check where the decision-maker asks:

- Why am I doing this?
- What do I think will happen?
- Is the benefit proportional to the risks?

The purpose of the decision control process is to guard against decision traps that may occur as a result of their management of the situation. Their research has found that applying these rapid checks improved situational awareness to a significantly higher level by supporting "anticipation" and has also mitigated some of the problems of an automatic response. Importantly, when the decision control process was used, doing so did not take unnecessary time away from executing the plan.

In a similar way, Judith Orasanu talks about asking "What's the problem?", "How much time is available?", and "How risky?". She also asks, "What makes this a difficult decision?". This includes ambiguity, conflicting cues, etc.

One strategy decision-makers can use is what Dirk McLean calls developing "meta-recognition" which is the ability to decide what is the appropriate way to arrive at a decision (Maclean, 2017). These include consideration of when the benefits of taking time to think critically are outweighed by the costs. For example:

- when the costs of delay are acceptable (i.e., there is time available for critical thinking)
- the situation is uncertain or novel (the recognized conclusions are subject to improvement)
- the costs of an error in acting on the current recognitional conclusion are high.

Another approach that helps is to become mindful of and better understand how cognitive bias operates to both help and hinder decision-making.

COGNITIVE BIASES AND BLIND SPOTS

Another word for cognitive bias is "heuristic", which means short-cut. For reasons explained in Chapter 3, our brains rely on shortcuts to conserve energy. This means we can subconsciously use a shortcut to reach a decision as effortlessly as possible. However, if the shortcut happens to be a blind spot, then this undermines our decision-making effectiveness.

The research by Cohen-Hatton et al. (2015) demonstrates that our challenge as decision-makers is to make quick decisions but also to be mindful of blind spots, biases, and other traps that can get in the way of making decisions, particularly in contexts of high consequence and under conditions of ambiguity and uncertainty. The brain also has a habit of applying a confirmation bias to prove it is right. That is, we naturally look for confirming evidence and others with similar explanations. Once the brain has an explanation, it can be difficult to move to a different perspective (the brain likes to be right, not wrong). Replacing the original explanation can take considerable effort. This is why developing sufficient storylines or narratives to explain an emerging situation is important (see Chapter 11).

Know Thy Biases

One way of avoiding error traps is to recognize the kinds of biases that exist and are likely to be prominent when we are under pressure. The good news is that we can easily identify several of the main ones presented in the Table below. The bad news is that, unfortunately, knowing about biases as error traps does not make us immune from falling into them. In fact, studies reported in Duke (2018) found that even if people were aware of their own biases, they were not better at being able to overcome them. However, with practice we can more easily recognize the potential for biases as blind spots in the thinking of others. This is why creating conditions of psychological safety becomes important in team effectiveness, discussed in more detail in Chapters 8 and 9. A list of some of the biases that can affect your thinking is included in the Table 2.1.

The challenge in VUCA environments for decision-makers is that even if a decision is perfectly made, the outcome may still be sub-optimal for a range of situational factors. This adds to the stress or pressure such decision-makers are under.

TABLE 2.1
Typical Cognitive Biases

Biases	Types
Attentional	*Framing:* The same problem receives different responses depending on how it is described.
	Anchoring: Give excessive weight to an unimportant but salient feature of the problem. Also associated with the relativity trap (we tend to pick the middle option).
	Availability or current moment bias: Captured by the current moment or the most recent thing discussed.
	Sunk cost: Tendency to continue with a plan or action because of the effort that has already been invested, even if it is not working.
Relational	*Bandwagon effect:* Go with the crowd (groupthink).
	False consensus: Assume most people think like us.
	In-group/out-group: Tendency to accept opinions from people who are more like us and to distrust information from others.
	Halo/devil effect: Stereotyping people based on previous experiences.

(Continued)

TABLE 2.1 ***(Continued)***
Typical Cognitive Biases

Biases	Types
Projectional	*Self-serving:* Desire for a positive self-image; tendency to claim more responsibility for successes than failures.
	Egocentric bias: Avoidance of unpleasant cognitive dissonance (i.e., ignore our own weaknesses).
	Fundamental attribution error: Over-emphasize personality-based explanations and under-emphasize the role and power of situational influences.
	Overconfidence bias: A tendency to over-estimate the effectiveness of our own strategies and under-estimate the effectiveness of the enemy (connects to planning fallacy – that our plans will always work).
Evidence-based	*Hindsight bias:* Inclination to see past events as being predictable.
	Confirmation bias: Tendency to search for or interpret information in a way that confirms one's preconceptions.
	Gambler's fallacy: We tend to put a tremendous amount of weight on previous events, believing that they'll somehow influence future outcomes (e.g., coin tossing).

METACOGNITION

Practitioners operating under pressure are required to manage competing cognitive demands so they remain focused on the critical issues (Frye & Wearing, 2014). Metacognition refers to thinking about and managing our thought processes (Frye & Wearing, 2014). Research with fireground commanders found that under-pressure decision-makers are vulnerable to two main types of error: fixation errors and fibrillation errors (Frye & Wearing, 2014). As the name suggests, fixation errors are a form of tunnel vision, where an over-focus on one aspect of the event leads to blind spots regarding other aspects of the event. This fixation can mean that important cues, albeit often subtle, may be missed, leaving not enough time to formulate and implement decisions for the changed circumstances. Fibrillation errors are almost the opposite of fixation errors and occur when people change their focus so frequently that they are unable to make progress on any critical tasks.

When decision-makers are caught in these two error traps (fixation and fibrillation), cues may not be detected, leading to blind spots. Changes may not be picked up in sufficient time. In their research, Frye and Wearing (2014) found that commanders were able to mitigate the potential for these error traps by engaging in metacognitive practices.

These experienced commanders used what Frye and Wearing (2014) call a metacognitive loop of monitoring the situation, deciding, and acting. Experienced commanders monitored where their attention was focused (is it on the big picture or in the detail and has it been there too long?). In thinking about their decisions, they monitored the tempo of those decisions. In this way, they were mindful of the balance between the need for speed and the need for accuracy. They considered their decision strategies (am I being defensive or offensive and are the conditions still appropriate for this action or have they changed, meaning my decision choice needs to change?). Finally, they monitored whether they were trying to do too much themselves and needed to bring in extra resources.

Frye and Wearing (2014) propose that practitioners can develop cognitive flexibility by being mindful of where their attention is focused (e.g., being able to move between the big picture and a more detailed view of different aspects of the situation or event). Decision-makers need to monitor their thinking and to calibrate these different stances.

STRATEGIES FOR PRACTITIONERS

It is helpful to understand where, as a decision-maker, your sources of pressure come from. Doing so enables you to anticipate ahead of time the areas where you are likely to be compromised when you are in the "hot seat".

There are many sources of indirect pressure, and these influence your capacity to think clearly and make decisions. Tool 2.1 contains two parts of a self-assessment tool to examine and identify sources of pressure and where they come from. The idea here is that if you can identify potential sources of pressure in your work environment ahead of time, you can do something about addressing them to mitigate their pressure in the future.

TOOL 2.1 ASSESSING SOURCES OF PRESSURE AT WORK

To use the tool, reflect on the items included in 2.2.1, thinking about a specific situation or place where you work and assess the degree of pressure you felt for each of the statements. Then insert the equivalent ticked column number for each of the statements ticked into Tool 2.2.2. For example, four of the items in Tool 2.2.2 indicate pressure for "role ambiguity" (items 2, 6, 7, and 9). If you felt no pressure for item 2 "Roles and duties of others I was working with were unclear to me", then you probably ticked the 1st or 2nd column, which would be a strength rating score of 1 or a 2. If item 6 "I was not completely sure what was expected of me" was a great source of pressure for you, then you probably ticked the 6th or the 7th column, and so on for statements 7 and 9. Then you would have four numbers to add up to get a total score for "role ambiguity".

Once completed, compare your scores for the four sources of pressure "role ambiguity", "role conflict", "role overload", and "responsibility for other people". Which one is higher? If the scores for the pressure indicator are

- between 4 and 10 [= low pressure]
- between 11–19 [= moderate pressure]
- between 20–28 [= high pressure].

TOOL 2.2.1
Identifying sources of pressure at work

2.2.1 Think about a time when you have been working under pressure and responsible for others. On a scale of 1 to 7, how strongly did these elements contribute to a feeling of being under pressure? (please tick)

		Little pressure ⟷ High pressure							
		1	2	3	4	5	6	7	N/A
1	I was responsible for helping others solve their problems	○	○	○	○	○	○	○	○
2	Roles and duties of others I was working with were unclear to me	○	○	○	○	○	○	○	○
3	I was accountable for the health and well-being/safety of other members of my team	○	○	○	○	○	○	○	○
4	At times the demands placed on personnel appeared unreasonable	○	○	○	○	○	○	○	○
5	I was held accountable for the work of others	○	○	○	○	○	○	○	○
6	I was not completely sure what was expected of me	○	○	○	○	○	○	○	○
7	At times I was unsure of my authority to carry out my responsibilities	○	○	○	○	○	○	○	○
8	The tasks I was undertaking seemed to be getting more complex	○	○	○	○	○	○	○	○
9	At times I was unclear about to whom I reported and/or who was reporting to me	○	○	○	○	○	○	○	○
10	The formal lines of authority were sometimes not adhered to	○	○	○	○	○	○	○	○
11	The actions and decisions I was taking impacted on the safety or well-being of others	○	○	○	○	○	○	○	○
12	The role I was doing was accepted by some but not by others	○	○	○	○	○	○	○	○
13	I sometimes received conflicting requests from two or more people	○	○	○	○	○	○	○	○
14	I had more work than can realistically be done in any shift	○	○	○	○	○	○	○	○

(Continued)

		Little pressure ⟷ High pressure							
		1	2	3	4	5	6	7	N/A
15	I got caught in the middle between my superiors and my subordinates	○	○	○	○	○	○	○	○
16	The tasks to be achieved were sometimes too difficult or too complex	○	○	○	○	○	○	○	○

Source: Fields (2002).

TOOL 2.2.2

Combining sources of pressure into categories

Categories	Item number and rating scores				Totals
Role ambiguity	2	6	7	9	
strength rating	□	□	□	□	□
Role conflict	10	12	13	15	
strength rating	□	□	□	□	□
Role overload	4	8	14	16	
strength rating	□	□	□	□	□
Responsibility for other people	1	3	5	11	
strength rating	□	□	□	□	□

Source: Fields (2002).

MAIN POINTS

- Decision-making in VUCA (volatile, uncertain, complex, and ambiguous) environments is challenging due to incomplete information and time constraints.
- The stress from high-pressure situations can trigger a fear response, leading to reactive and short-term focused decisions.

- Two primary thinking styles are used in decision-making: rational (analytical) and intuitive (sub-conscious).
- Understanding these thinking styles is essential for improving decision-making capabilities under pressure.
- Common pitfalls in decision-making include biases such as anchoring, availability bias, and the sunk cost fallacy.
- Strategies to enhance decision-making include metacognition, decision control processes, and situational awareness techniques.
- A tool has been provided to assist in identifying sources of pressure at work help assess role ambiguity, conflict, overload, and responsibility for the well-being of others, which can impact decision-making.

REFERENCES

Butler, P. C., Bowers, A., Smith, A. P., Cohen-Hatton, S. R., & Honey, R. C. (2023). Decision making within and outside standard operating procedures: Paradoxical use of operational discretion in firefighters. *Human Factors*, *65*(7), 1422–1434. https://doi.org/10.1177/00187208211041860

Cohen-Hatton, S. R., Butler, P. C., & Honey, R. C. (2015). An investigation of operational decision making in situ: Incident command in the UK fire and rescue service. *Human Factors*, *57*(5), 793–804.

Duke, A. (2018). *Thinking in bets: Making smarter decisions when you don't have all the facts*. Portfolio Press.

Fields, D. L. (2002). *Taking the measure of work: A guide to validated scales for organizational research and diagnosis*. Sage.

Frye, L., & Wearing, A. (2014). What were they thinking? How the experts make decisions during a large-scale bushfire. In C. Owen (Ed.), *Human factors challenges in emergency management*. Ashgate Press.

Kahneman, D. (2011). *Thinking fast and slow*. Farrar, Straus and Giroux.

Klein, G. (1997a). Developing expertise in decision making. *Thinking and Reasoning*, *3*(4), 337–352.

Klein, G. (1997b). The recognition-primed decision (RPD) model: Looking back, looking forward. In C. Zsambok, & G. Klein (Eds.), *Naturalistic decision making*. Lawrence Erlbaum Associates.

Klein, G. (2011). Critical thoughts about critical thinking. *Theoretical Issues in Ergonomics Science*, *12*(3), 210–224. https://doi:10.1080/1464536x.2011.564485

Maclean, D. (2017). Shoot, don't shoot: Minimising risk of catastrophic error through high consequence decision making. Air Power Development Centre.

Molins, F., Ayuso, C., & Serrano, M. (2021). Emotional stress & decision-making: An emotional stressor significantly reduces loss aversion. *Stress*, *24*(6), 780–786.

Orasanu, J., & Connolly, T. (1993). The reinvention of decision making. In G. Klein, J. Orasanu, R. Calderwood, & C. E. Zsambok (Eds.), *Decision making in action: Models and methods* (pp. 3–20). Ablex.

Penney, G., Launder, D., Cuthbertson, J., & Thompson, M. B. (2022). Threat assessment, sense making, and critical decision-making in police, military, ambulance, and fire services. *Cognition, Technology & Work*, *24*(3), 423–439. https://doi.org/10.1007/s10111-022-00694-3

Sarmiento, L. F., da Cunha, P. L., Tabares, S., Tafet, G., & Gouveia, A. Jr. (2024). Decision-making under stress: A psychological and neurobiological integrative model. *Brain, Behavior, & Immunity-Health*, *38*, 100766.

Starcke, K., & Brand, M. (2012). Decision making under stress: A selective review. *Neuroscience & Biobehavioral Reviews*, *36*(4), 1228–1248.

't Hart, P. (2014). Managing crises long term: The challenges of preparedness and response. Cited in Eburn, M., & Gissing, A. (2024). Principles of best practice strategic crisis management arrangements for catastrophic disasters. Natural Hazards Research Australia.

Tversky, A., & Kahneman, D. (1974). Judgment under uncertainty: Heuristics and biases. *Science*, *185*, 1124–1131.

3 Decision-Making and Stress

INTRODUCTION

As discussed in the previous chapter *Making Decisions Under Pressure*, when decision-makers are faced with VUCA (volatility, uncertainty, complexity, and ambiguity) related conditions (see Chapter 1), they are typically under pressured constraints, and these contribute to increased stress. Understanding the link between decision-making and stress is important as it underpins actions necessary for survival (Sarmiento et al., 2024). As a predictive organ, the brain creates models of what it expects to occur based on previous experiences and uses these to anticipate future events (Šimić et al., 2021). When events do not follow the patterns predicted, we can experience stress psychologically and physiologically in relation to the degree or otherwise of perceived threat. This can happen consciously and unconsciously through the activation of a set of nuclei in the base of the brain called the amygdala (Šimić et al., 2021).

When we are faced with a perceived threat, our body reacts by increasing our heart rate, mobilizing glucose, and centralizing blood circulation in preparation for maximum activation. At the time of imminent danger, this is helpful. A surge of adrenaline provides the energy needed for concentration and alertness (Ascher & Tonies, 2021). Under these conditions, stress can serve as a motivating force, enabling us to focus and energize ourselves for enhanced performance.

However, stress can also manifest negatively, leading to distress. Distress occurs when individuals find themselves unable to cope with or adapt to the challenges or situations they are facing (Australian Red Cross, 2020), resulting in physical and emotional discomfort. This can stem from a singular crisis event or accumulate over time. In acute stress, studies have shown that we are more likely to respond with well-learned habits and routines to optimize rapid responses (Sarmiento et al., 2024). Our actions shift from thoughtful to more rapid and reflexive (Ressler, 2010). We are likely to experience challenges using our short-term memory for problem solving, though conversely, we are more likely to establish long-term memories that encourage learning from fear-inducing events (Sarmiento et al., 2024).

STRESS AND DECISION-MAKING

For personnel working in high-stress occupations (e.g., police, military, ambulance, critical care), stress is inherent in one's occupation. These include shift work, work-related violence, overtime demands, and traumatic events, whereas the context-based

DOI: 10.1201/9781003566007-4

stressors are likely to be the interaction with managing the event and its potentially negative outcomes.

Stress impacts on the quality of our thinking. Most of the time, we don't have a lot of working memory (typically seven plus or minus two new units of information), though when we are stressed, our short-term working memory reduces considerably. This has implications for the training of responders who, according to McLennan et al. (2014), may have difficulty remembering procedures or in correctly attending to and interpreting the importance of emerging threats. Our capacity for abstract reasoning is also depleted when stressed, which means we might not be able to think rationally or apply logic in ways that can efficiently consider options. In addition to the impact of stress on thinking, stress also has negative impacts on physical activities. McLennan et al. (2014) found that movements can be slower and less skillful, and we can sometimes take longer to complete tasks (e.g., deploying fire shelters).

Research evidence suggests (e.g., Dilawar et al., 2021) that under stressful conditions a rational analytical decision-making approach is less likely to be used. The argument is that under conditions of time pressure, insufficient information, and high consequences (e.g., emergencies) personnel cannot always analyze all contingencies and instead they rely on intuitive means and experience of previous events.

There is also a strong body of evidence (e.g., Dilawar et al., 2021; Hallo & Nguyen, 2021) that suggests decision-makers operating in dynamic (i.e., rapidly shifting) environments (see Chapter 2) will use both rational and intuitive forms of thinking and do so in ways that are complementary. From this perspective, rational and intuitive thinking styles work as parallel cognitive systems. However, as highlighted in Chapter 2 on cognitive bias, there are dangers and traps in doing so.

Working under pressure can bring with it strong emotional and physical distress. As a decision-maker, you need to have good skills at both taking care of yourself and others. You need to be able to optimize your performance and the performance of others. This involves understanding the impacts of stress and how you can optimize the role of stress in performance. In addition, research has identified that people high in emotional intelligence (discussed later in Chapter 5) are more capable of combating stress than those with low or moderate emotional intelligence (Dilawar et al., 2021). However, there is also some evidence (Nitschke & Bartz, 2023) that under conditions of acute stress, we can lose affective empathy which may in turn undermine decision-making and leadership effectiveness (see Chapter 7).

Understanding Stress and Styles of Coping

There are typically three styles decision-makers use to cope with stress, and one is not better than the other. It is important, however, to identify which style is one you are more likely to use. Being self-aware and able to reflect on the strengths and weaknesses of your approach is the key.

Avoidant style of coping – rather than focus on the stress in the moment these decision-makers focus on the task that needs to be done. They feel the emotion afterwards.

Ruminative style – these decision-makers do feel the emotion and stress in the moment – they think about it and deal with it in the moment and then afterward continue to ruminate on it.

Hyper vigilant – when these decision-makers experience high stress – these leaders feel the tension and the emotion. They are in the moment based on the physiological reaction they make decisions very quickly.

Each style has strengths and blind spots. To avoid burnout, it is important for each to address the pressure that has built up. For those with an avoidant style of coping, it is important to have regular down time. The brain learns this is part of a daily routine and very quickly shifts to recovery mode because it knows that it has that downtime where it is not going to be thrust back into work mode. Taking a step back, reading books, gardening, and allowing your mind to reflect on what stress has been experienced are important. An avoidant style needs to reconnect with experiences in order to process the stress. For those with a ruminative style, the above activities are also useful. Down time is particularly important in order to focus attention away from the things you ruminate on – it is important to allow your brain to change the channel. For those with a hypervigilant response having an exercise routine as part of your daily regime is particularly important to get back to baseline.

The Amygdala Hijack

The amygdala, which acts as the brain's emotional center, is highly important for survival. Daniel Goleman (1995) first coined the term "amygdala hijack" to describe what can happen when this part of the brain becomes highly activated in a significant threat state (see also Chapter 5 for more discussion on Goleman's work). The amygdala can override all other thinking leading to an overwhelming emotional response that may be either inappropriate or disproportionate to the situation. Zaria Gorvett (2017) described inappropriate actions that can be characterized by an amygdala hijack. These include, in addition to a fight-flight response, one known as freezing. For example, in the 2017 terrorist attack on Westminster Bridge where the perpetrator drove a car into pedestrians, injuring more than 50 people, first responders described members of the public standing "like deer in the headlights" (Gorvett, 2017). In a nutshell, in the face of danger, we can become immobilized. In addition to doing nothing, sometimes people become stuck in a routine of doing the same thing over and over again, which is described by Gorvett (2017) and others as "perseveration".

If the amygdala is in overload, our emotional responses can result in disproportionate outbursts where we are not perceiving or assessing information accurately or able to monitor external feedback properly. Strategies to counter an amygdala hijack include slowing down and refocusing, breathing (see Chapter 4), and becoming more mindful of the triggers though emotional self-awareness (see Chapter 5), including using self-reflective activities such as those described at the end of this chapter.

It is also important to acknowledge that the amygdala plays a role not only in regulating fear responses from external stimuli but also from learned fear. This is the reason why the amygdala has been implicated in post-traumatic stress disorder (Ressler, 2010). In chronic stress, the amygdala remains highly activated thereby increasing fear and anxiety responses. In post-traumatic stress, these responses are both enhanced and are difficult to be extinguished or inhibited (Davis & Hamner, 2024).

Pearlman and McKay (2025) identify that people suffering vicarious trauma tend to struggle with one or more of the following:

- Moral injury: Tragic critical incidents weaken the sense of hope, purpose, and spiritual empowerment necessary to help victims of disasters and emergencies.
- Physical injury: Repeated exposure to toxins, physical stress, accidents, and danger reduces fitness, comfort, and overall well-being.
- Emotional injury: Acute traumatic stress and chronic emotional stress increase the possibility of anxiety, depression, and PTSD.

Self-Awareness and Understanding Burnout

Table 3.1 has been adapted from McNaughton et al. (2015). Review these items for yourself or a colleague. The greater the number of ticks, the higher likelihood of burnout.

This burnout checklist is not clinical or definitive and should be used as an indicator only. Even if the items on this checklist are not resonating with you, if you're

TABLE 3.1
Understanding Burnout

Item	Tick If Present
Starting to resent clients or becoming overly involved and taking on their worries	O
Lacking enthusiasm, drive, and energy	O
Experiencing a loss of capacity for empathy	O
Feelings of emotional withdrawal, bitterness, and cynicism	O
Having a reduced sense of accomplishment	O
Developing a narrow focus and inability to see where your role fits in the "bigger picture"	O
Tending toward poor or uncooperative behavior to spite "the system"	O
Being overwhelmed by the size of the task, and defensive about expectations	O
Feeling challenged by change; resorting to responses such as "This is how we do it!"	O
Over time becoming too invested and unable to accept when your role is no longer required	O
Feeling that you are indispensable to a community or individual's recovery	O

Source: McNaughton et al. (2015).

concerned for your wellbeing, seek advice and support. Dr Laurie Pearlman (2012) of the Headington Institute suggests 3 strategies for self-care. These include

- Become aware of what's getting to you and how those issues are affecting you
- Seek balance among work, rest, and play; time alone and time with others; giving and receiving
- Connect with people (friends, family, co-workers) you trust, respect, and care about.

Chronic Stress and Allostatic Load

Research has shown that stress can drain emotional resources (Grubb et al., 2018). Challenges arise when individuals do not adequately recover from stress and start to internalize it unconsciously. As discussed, stress can be useful for extra energy and focus – but when it is continuous, it begins to change our brains. Chronic stress opens the door to burnout and post-traumatic stress disorder.

Self-awareness is fundamental to effectively managing stress. Table 3.2 describes some of the sources of chronic stress and the ways in which we can begin to absorb stress unconsciously, especially after a stressful episode.

To harness stress as a beneficial force and maximize its advantages, we must consciously engage in energizing activities while also prioritizing recovery to process our experiences and rest. Through practice, we can learn to move deliberately between an engaged state, where we are energized, focused creative, and productive, and a recovery state where our brain processes events, learns, and recuperates.

Resilient people operating in crisis are aware of these stressors and develop strategies to turn off what is known as the allostatic load and bring us back to a relaxed state. Allostatic load is the outcome of cumulative effects of chronic stress and the impact of that stress on our physical and mental health (see for example, Guidi et al., 2021). It is also important to get serious about sleep and to be eating well, keeping hydrated, and also attending to breathing (discussed more detail in Chapter 4).

TABLE 3.2
Warning Signs of Chronic Stress

Physical	Behavioral	Emotional	Cognitive
Headache	Bossiness/arguing	Anxiousness	Not thinking clearly
Stomachache	Increased alcohol/substance use	Boredom	Forgetfulness
Muscle tightness	Compulsive eating/smoking	Edginess	Indecision
Elevated heartrate	Shouting/crying/withdrawal	Powerlessness	Inability to focus

Source: Caillet et al. (2016).

Note: Chronic stress can lead to burnout which the World Health Organization classified, in 2019, as an occupational phenomenon characterized by three dimensions: feelings of energy depletion or exhaustion; increased mental distance from one's job, or feelings of negativism or cynicism relating to one's job; and reduced professional efficacy.

RESILIENCE IN DISASTERS

According to Stephen Thompson at the Headington Institute (2025), practitioners who maintain their well-being have been identified as having developed the following resilience elements:

- Social support: Make regular connections with friends and family.
- Self-efficacy: Develop confidence and competence in your own self-efficacy, particularly in recognizing and accessing your own emotional states.
- Meaning and purpose: Clarify your own sense of vocation.
- Physical fitness: Exercise regularly to maintain your health.

SELF-CARE

There are a variety of activities that all people involved in making decisions under pressure who suffer stress can do to support self-care. The personal recovery rocket (see Figure 3.1) highlights some of these activities and encourages people to try to participate in a mix of these across weeks, months, and years. For example, for 30 weeks of the year, you are encouraged to tally up 100 recovery points from activities such as those in Table 3.3. The personal recovery rocket might be helpful for you to look over when developing your own self-care plan.

STRATEGIES FOR PRACTITIONERS

The following strategies can assist in better managing stress when under pressure.

Locating and describing your stress is a process of discovery, and the more you learn, the better you will harness your own stress response. Tool 3.1, adapted from

FIGURE 3.1 Personal Recovery Rocket. (From May, A. (2010). *Recovery Rocket*. Retrieved April 28, 2025, from https://performanceintelligence.com/blog/)

TABLE 3.3
Recovery Points

Indoor Activity	Points	Outdoor Activity	Points
Massage	50	Relaxing swim	25
Meditation	50	Surf or paddle	25
Stretch, yoga, tai chi	40	Bushwalk	25
Visualization	30	Barbeque/picnic	25
Diaphragmatic breathing	30	Golf	25
Warm bath	30	Fishing	25
Social catch up	20	Gardening	25
Church/spiritual service	25	Slow walk	20
Movie or theatre	25	Slow cycle	20
Relaxing music	15	Park with kids	20
Read fiction	15	Sightseeing or festival	15
TV	5	Shopping	15

Source: May (2010).

Ascher and Tonies (2021), has been devised to increase awareness of your stress, and help to manage it.

QUESTIONS FOR REFLECTION

Rest and recovery: In a typical week, how often do you feel rested? During your shifts, do you intentionally seek out moments of recovery, or do you tend to "power through"?

Engagement in work: How often during a typical shift do you feel fully immersed in your tasks?

STRATEGIES TO REDUCE AND MANAGE STRESS

Become mindful of your stress: Acknowledge your stress levels and identify triggers.

Transition awareness: Consciously focus on the shifts between engaging in stressful tasks and entering recovery periods. This might involve simple actions like taking a walk or practicing deep breathing exercises.

Staying present: One of the hallmarks of elite performance in sports is the ability of athletes to reset quickly after making mistakes. Instead of dwelling on errors, they concentrate on the present moment, which enhances their focus and resilience. Tool 3.3.1 is a grounding exercise designed to manage acute stress, reduce anxiety, and readjust focus. It is adapted from the work of Esther Yu of the Headington Institute.

TOOL 3.1 PRACTICE COUNTING DOWN TO CALM

This is known as the 5-4-3-2-1 Technique. Its purpose is to help bring the mind back to the present moment and to help ground us rather than becoming anxious or flighty.

TOOL 3.3.1
The 5-4-3-2-1 technique

5	Acknowledge FIVE things you SEE around you. Maybe it is a bird, something on your desk, or the color of a particular piece of furniture. However big or small, state 5 things you see.
4	Acknowledge FOUR things you can TOUCH around you. This could be your hair, hands, the grass, the ground, or the chair you're sitting on. Whatever it may be, list 4 things you can touch.
3	Acknowledge THREE things you can HEAR. These should be external, do not focus on your thoughts. Maybe you can hear a car, the air conditioning, or a nature sound. Focus on things you can hear outside of your body.
2	Acknowledge TWO things you can SMELL. This one might be harder if you're not in a stimulating environment. If you cannot automatically sniff something out, walk nearby to find a scent. Maybe it is the smell of paper, or a pillow, or something from nature outside.
1	Acknowledge ONE thing you can TASTE. What does the inside of your mouth taste like? Gum, coffee, tea, whatever you had for lunch? If this one is difficult, name something you could taste around you, such as food that you see on the counter.

Source: Yu (2020).

Note: Taking these steps will not erase your stress or anxiety overnight, but it can be a very handy way of coping and significantly reducing the intensity of these experiences.

MAIN POINTS

- Practitioners operating in stressful environments will face challenges with recall and are more likely to engage in intuitive thinking. Stress can also negatively impact on physical activities too.
- Understanding the changes in the brain when stressed, such as the negative impacts caused by an amygdala hijack, is important to self-awareness.
- Increasing self-awareness about chronic stress and allostatic load is important in mitigating burnout and posttraumatic stress disorder.
- Chronic stress can lead to burnout, characterized by energy depletion, mental distance from work, and reduced professional efficacy.
- To effectively manage stress, individuals should alternate between engaging in energizing activities and prioritizing recovery to process experiences.

- Some strategies for self-care are included as well as a personal recovery rocket to monitor rest activities.
- A tool is provided to support a grounding exercise that can help decision-makers stay present and manage acute stress by focusing on the current moment.

REFERENCES

Ascher, J., & Tonies, F. (2021). How to turn everyday stress into 'optimal stress'. McKinsey Quarterly. Retrieved April 10, 2025, from https://www.mckinsey.com/business-functions/people-and-organizational-performance/our-insights/how-to-turn-everyday-stress-into-optimal-stress.

Australian Red Cross (2020). Psychological first aid. Retrieved April 28, 2025, from https://www.redcross.org.au/globalassets/cms-assets/documents/stories/psychological-first-aid-an-australian-guide.pdf.

Caillet, A., Hirshberg, J., & Petti, S. (April 15, 2016). *A simple way to combat stress.* Harvard Business Review. World Health Organization.

Davis, L. L., & Hamner, M. B. (2024). Post-traumatic stress disorder: The role of the amygdala and potential therapeutic interventions – A review. *Frontiers in Psychiatry, 15*, 1356563.

Dilawar, S. M., Durrani, D. K., Li, X., & Anjum, M. A. (2021). Decision-making in highly stressful emergencies: The interactive effects of trait emotional intelligence. *Curr. Psychol, 40*(6), 2988–3005. https://doi.org/10.1007/s12144-019-00231-y

Goleman, D. P. (1995). *Emotional intelligence: Why it can matter more than IQ.* Bantam Books.

Gorvett, Z. (2017). *What not to do in a disaster.* Retrieved April 28, 2025, from https://www.bbc.com/future/article/20170711-what-not-to-do-in-a-disaster

Grubb, A. R., Brown, S., & Hall, P. (2018). The emotionally intelligent officer? Exploring decision-making style and emotional intelligence in hostage and crisis negotiators and non-negotiator-trained police officers. *Journal of Police and Criminal Psychology, 33*(2), 123–136. https://doi.org/10.1007/s11896-017-9240-2

Guidi, J., Lucente, M., Sonino, N., & Fava, G. A. (2021). Allostatic load and its impact on health: A systematic review. *Psychotherapy and Psychosomatics*, 90(1), 11–27. https://doi.org/10.1159/000510696.

Hallo, L., & Nguyen, T. (2021). Holistic view of intuition and analysis in leadership decision-making and problem-solving. *Administrative Sciences, 12*(1), 4.

May, A. (2010). *Recovery Rocket.* Retrieved April 28, 2025, from https://performanceintelligence.com/blog/

McLennan, J., Strickland, R., Omodei, M., & Suss, J. (2014). Stress and wildland firefighter safety-related decisions and actions. In C. Owen (Ed.), *Human factors challenges in emergency management: Enhancing individual and team performance in fire and emergency services.* Ashgate Publishing.

McNaughton, E., Wills, J., & Lallemant, D. (2015). *Leading in disaster recovery: A companion through the chaos.* Global Disaster Preparedness Center. https://preparecenter.org/wp-content/sites/default/files/leading_in_disaster_recovery_a_companion_through_the_chaos.pdf

Nitschke, J. P., & Bartz, J. A. (2023). The association between acute stress & empathy: A systematic literature review. *Neuroscience & Biobehavioral Reviews, 144*, 105003.

Pearlman, L. (2012). *What to do about burnout: Identifying your sources.* Headington Institute. https://headington-institute.org/wp-content/uploads/2020/02/R3-What-To-Do-About-Burnout.pdf

Pearlman, L., & McKay, L. (2025). *Understanding and addressing vicarious trauma*. Online training module 4: Headington Institute, CA. Retrieved April 28, 2025, from https://headington-institute.org/wp-content/uploads/2024/10/Module-Four-Understanding-Addressing-Vicarious-Trauma-Reading-Course-Module-Four.pdf

Ressler, K. J. (2010). Amygdala activity, fear, and anxiety: Modulation by stress. *Biol. Psychiatry*, *67*(12), 1117–1119. https://doi:10.1016/j.biopsych.2010.04.027

Sarmiento, L. F., da Cunha, P. L., Tabares, S., Tafet, G., & Gouveia, A. Jr. (2024). Decision-making under stress: A psychological and neurobiological integrative model. *Brain, Behavior, & Immunity-Health*, *38*, 100766.

Šimić, G., Tkalčić, M., Vukić, V., Mulc, D., Španić, E., Šagud, M., Olucha-Bordonau, F., Vukšić, M., & Hof, P. R. (2021). Understanding emotions: Origins and roles of the amygdala. *Biomolecules*, *11*(6), 823. https://doi.org/10.3390/biom11060823

Thompson, S. (2025). *Self-compassion for wildfire responders: Managing guilt, compassion fatigue, and responsibility*. Headington Institute. Retrieved April 28, 2025, from https://headington-institute.org/blog/resource/self-compassion-for-wildfire-responders-managing-guilt-compassion-fatigue-and-responsibility/

Yu, E. (2020). *Countdown to calm: Using 5-4-3-2-1- to recenter and reground*. Headington Institute. www.headington-institute.org/resource/countdown-to-calm

4 Mind and Body

INTRODUCTION

As discussed in the previous chapter, stress makes making decisions more challenging and the primary organ involved is the amygdala, which, when activated, can lead to an overwhelming emotional response. While much research and attention have been given to the role of the brain in complex decision-making, it is important not to overlook the role of the body and of embodiment. Joas (1996) contended that most theories of action overlook the role of the body and are based on an implicit assumption that "the body is the factual basis of action but pay no attention to it, as if in a fit of theoretical prudishness" (p. 167). Recent research has highlighted the increasing appreciation of how thinking and decision-making does not reside merely within our heads (Bechara & Damasio, 2005; Lucas, 2018). Owen (2024) highlighted how in high-intensity, high-tempo, high-consequence work, the body is used as a resource to monitor the temporal flow of the work and bodily signals to monitor their management of the intensity so a practitioner knows in advance when they are about to be overwhelmed and need to seek help.

This chapter explores the gut-brain connection to better understand a range of other bodily functions and how they, too, are communicating with our brains and thus influencing decision-making when under pressure. The strategies we can use with our bodies will then be discussed.

GUT AND BRAIN IN DECISION-MAKING

The brain and the gut are highly interconnected. Indeed, researchers at *Johns Hopkins University School of Medicine* claim the gut (or the enteric nervous system) is a second brain. According to Ferguson (2023), the vagus nerve, which connects the enteric nervous system to the brain, acts as an information superhighway. What is interesting is that it is not the brain sending most of the information to the body but the other way around. In fact, the body sends four times as much information to the brain as the brain sends to the body. That is, 80 percent of the information on the superhighway goes from your body to your brain, and only 20 percent of the information comes from your brain to your body.

The body does this through a process called neuroception to process information about our surroundings (safe or dangerous?) and tracks how other bodily functions are working, though most of the time we are unconscious of it. In short, what this implies is that while the brain uses the central nervous system to affect the gut, there is also direct communication from the gut and the gut microbiota to the central nervous system, allowing the gut to influence brain activity and behavior (O'Mahony et al., 2015).

DOI: 10.1201/9781003566007-5

TABLE 4.1
Five Simple Ways to Improve Gut Health

Adding fiber to your diet	Adding more fiber to your diet helps gut health because it serves as a prebiotic, that is, food for the good bacteria in your gut. Fiber is found in foods such as legumes, whole grains, avocados, sweet potatoes, Brussels sprouts, tofu, berries, leafy greens, and nuts and seeds.
Staying hydrated	Try to drink between four and six glasses of water a day. Water is important for healthy digestion.
Managing stress	When the hormones, discussed in Chapter 3, spike, it can lead to digestive discomfort. The strategies included later in this chapter can also assist in managing stress.
Getting sufficient sleep	See the suggestions for practitioners to support good sleep later in this chapter.
Getting psychically active	Exercising between 150 and 270 minutes a week has a positive effect on your gut microbiota.

Source: Warner (2025).

Research has explored the links between digestion, mood, and cognition (thinking skills and memory). Lucas (2018), for example, found that the quality of our intestinal microbes has a significant influence on cognition and mood. There is an increasing body of evidence to suggest that the gut microbiome plays a significant modulatory role in a range of social cognition and behavior (Falkenstein et al., 2024). In a seven-week experimental study, Falkenstein et al. (2024) manipulated the gut microbiome in 101 male participants and compared the results of the 2 randomized (experimental/placebo) groups. The group who received dietary supplements increasing the diversity in their gut flora showed a significant improvement in a pro-social altruistic decision-making exercise. In a similar study examining decision-making involving risk-taking in time-pressured environments, Dantas et al. (2021) found a significant decrease in risky behavior and an increase in future-oriented choices in the probiotics group as compared to the placebo group.

In a nutshell, a proper ecological balance of gut microbiota has been found to affect cognitive performance, mood as well as susceptibility to stress (Dinan et al., 2015). In writing for the Harvard Health publishing unit, Warner (2025) identifies five simple ways to improve gut health (see Table 4.1).

EMBODIED EMOTION AND COGNITION

According to Lerner et al. (2015), the role of emotion in decision theories has led to a paradigm shift in the way we think about making decisions, particularly under pressure. This is, in part, because there is a reciprocal relationship between emotion and decision-making. Cognitive processes can shape emotion, and at the same time, emotion influences cognition. There is also a greater understanding now of how the body is a significant mediator of emotion.

In a review of the literature on emotion and its relationship between the body and the brain, Šimić et al. (2021) note that for a long time, it has been proposed that the body responds to the emergence of emotions. For example, tense muscles are interpreted by the brain as anxiety. Yet when the subject is given benzodiazepines, a drug that induces muscle relaxation, the brain no longer receives this information, and the subject becomes less anxious.

In the book *The Feeling of What Happens*, Damasio (1999) explains how emotions are unconsciously formed within the body's central nervous system and then only later are feelings consciously produced by our interpretation of the signals. However, as Šimić et al. (2021) also point out, once embodied, feelings can persist within the central nervous system as exemplified by phantom pain. Damasio's argument suggests that there is no pure perception without mediation of the body and the interpretation of bodily signals.

Damasio has also further developed what he calls the somatic marker theory of the connection between our bodies and stored emotions. His hypothesis suggests that emotions significantly influence decision-making processes and are central to cognition (Bechara & Damasio, 2005; Damasio, 1999). In essence, when practitioners face decisions – especially in intricate or uncertain contexts – they do not rely solely on rational thinking. Instead, they draw upon previously stored emotional experiences known as "somatic markers" (Bechara & Damasio, 2005). These markers consist of automatic associations linking different options to their emotional histories, reflecting positive or negative outcomes. This system allows practitioners to use these emotional cues as shortcuts, guiding them toward choices that have historically resulted in more favorable emotional experiences.

Categorization plays a vital role in this process, as it shapes the emotional significance of various situations. It enables individuals to utilize these concepts for emotional reasoning, decision-making, and effective communication with others. Affective categorization not only influences one's own emotional responses but also assists in recognizing and inferring the emotions of others through cues such as facial expressions, tone of voice, language, and other behaviors (Bechara & Damasio, 2005).

The implication here is that we can influence feeling and emotion by how we regulate our bodily motor behavior. Therapeutic application of this insight has contributed to, for example, clinicians using dance psychotherapy to facilitate processing emotions and their regulation through movement (Shafir et al., 2016).

Tool 4.1 can help practitioners to develop what Brassey et al. (2022) call dual-awareness. The techniques aim to interrupt the body's stress response while still being mindful of it.

Creating Transitions Between Focus and Rest

It is helpful to create transitions between engagements, and rest helps keep them separate and distinct in our brain. This in turn increases their effectiveness. The goal is to learn how to make the act of focus and recovery something you can call on when you need it as a part of your everyday. The seven types of rest, proposed by Dr Saundra Dalton-Smith (2017), are summarized in Table 4.2.

TOOL 4.1 DEVELOP MINDFUL BODY AWARENESS – PHYSICAL TECHNIQUES

Technique	Practice
Focus on the exhale	While breathing deeply is helpful to calm us down it is the exhale that is the most important. Deep breaths where the exhalation is longer than the inhalation decrease the heart rate. Try two quick inhales through the nose with a brief pause in between and then a long exhale through the mouth. Do this two or three times in sequence at the first sign of stress.
Zoom out	To counter the tunnel vision that accompanies the protective mode (and in which we are focused on what is comfortable and familiar) concentrate on your peripheral vision and take in a panoramic view of the environment. Focus on the bigger picture to the left and right as you see. This helps if it is practiced for a couple of minutes every day and also helps to lower stress levels in challenging moments.
Move	With the "amygdala hijack" (see Chapter 3), we are in a state of fight, flight or freeze. Instead of freezing and staying still, we can release tension by moving our bodies. A brisk five-to-ten-minute walk can convince the body that we are taking forward action. Doing so typically involves looking right and left thus taking in the panoramic view.
Calm your voice	Vocal techniques such as singing, humming chanting, and even gargling can help us feel calmer by activating the vagus nerve. Notice your voice and lower your pitch and volume, make a conscious effort to speak slowly.
Breathe deep	Focus on breathing from the belly and not the chest.

Source: Brassey et al. (2022).

TABLE 4.2
Seven Types of Rest

Type of Rest	Definition	Signs You May Need More of This Rest	Some Suggestions to Help Support This Type of Rest
Physical	Needed when the body is tired and needs some time to recover	Busy lifestyle or busy job Lacking physical energy Experience muscle soreness and weak immune system	Sleep/napping Stretching Breathing exercises Gentle walks Relaxing baths Massage
Mental	Allows space to declutter and calm the mind, slowing the pace	Mind racing, struggling with sleep Struggling to stay present and focused on one thing Easily irritated or frustrated Making lots of mistakes	Mindfulness and meditation Creating a safe space to switch off Getting lost in an easy-to-watch program or easy-to-read book

(Continued)

TABLE 4.2 ***(Continued)***
Seven Types of Rest

Type of Rest	Definition	Signs You May Need More of This Rest	Some Suggestions to Help Support This Type of Rest
Emotional	Need to give us space to check in with ourselves and rest from situations where we are holding the space for others	Working in a role with highly charged emotional spaces Increase in negative emotions – fear, anxiety, depression Focusing on negatives	Learn what refreshes you Check in with your emotions regularly throughout the day Become aware of what affects you Increase self-care Seek connection
Spiritual	A personal sanctuary. Where we check in and reclaim a sense of peace	Craving safety and stability Experiencing emotions of helplessness, numbness, or apathy Disconnected from your beliefs	Meditation Journaling Prayer Explore what is troubling you
Creative	Allow space for us to explore beauty and wonder	A job that requires a lot of creative thinking Struggle to see your needs or self-care as a priority Rarely feel like your work is of value Find it difficult to enjoy things of nature	Giving space for your natural response to being creative Art Music Nature Craft Dance
Sensory	Reducing over-stimulation of sense, we often experience	Sensory overload Strained vision/tired eyes Adverse reaction to loud sounds, bright lights Struggle with strong sensory experiences – fireworks, concerts, loud workplace	Unplug from technology at times 5 senses meditation – focus on one sense and tune out the others Identify your main sensory stressor and spend time away from it
Social	Reconnects us to people we love and gives time for rewarding relationships	A sense of social overload Feeling detached and lonely Struggle with face-to-face communication Craving time of being alone	Invest in the people who refresh and revive you Seek connection where you can be fully accepted

Source: Dalton-Smith (2017).

THE ROLE OF SLEEP IN SUPPORTING DECISION-MAKING

In considering the organizational cost of insufficient sleep, van Damn and van der Helm (2016) note that when we are sleep deprived, we lose the ability to focus attention and make accurate judgments. Indeed, they note that 17–19 hours of wakefulness are likely to lead to performances on tasks that are equivalent to someone with a blood-alcohol level of 0.05 percent and after roughly 20 hours of wakefulness, this will have risen on average to the equivalent blood-alcohol level of 0.1 percent.

We would not allow this person to drive a car, yet we allow them to be making decisions in high-consequence VUCA-related events. Evidence suggests that poor sleep patterns result in mood changes (increased anxiety or depression), memory issues (in both short- and long-term memory), and physical impacts such as a weakened immune system and higher blood pressure.

The impacts of sleep deprivation include

Thinking and concentration – concentration, creativity, and problem-solving skills aren't at their best when we don't get enough sleep. These skills are all important in interpreting and managing VUCA-related events.

Mood changes – sleep deprivation can make you moody, emotional, and quick-tempered. Chronic sleep deprivation can increase anxiety and depression. This can affect your capacity to lead your team.

Low sex drive – people who don't get enough sleep often have a lower libido. In men, this may be due to a drop in testosterone levels.

Memory issues – during sleep your brain forms connections that help you process and remember new information. A lack of sleep can negatively impact both short and long-term memory, which can have a dramatic effect on being able to lead in a crisis.

Weakened immunity – too little sleep weakens your immune system's defenses against viruses, likely those that cause the common cold or flu. You are more likely to get sick and be unable to lead.

Poor health – lack of sleep affects your body's release of insulin which leads to higher blood sugar levels and the increased risk for type 2 diabetes. There are also links to weight gain, high blood pressure, increased risk of heart disease and balance.

STRATEGIES FOR PRACTITIONERS

The following suggestions have been extracted from van Dam and van der Helm (2016) to enhance the quality of your sleep.

Create the Right Sleep Environment

- Remove devices such as smartphones, iPads, or computers from the bedroom.
- Don't use the bedroom for work.
- Keep the bedroom cool, allowing your core body temperature to drop, which helps you fall and stay asleep.
- If you travel for work, take items that remind your brain of home, such as your pillow, pajamas, shower gel, and toothpaste.

Wind Down

Quality of sleep is compromised when you don't sufficiently relax and reduce stress in the evening. It is critical to wind down at night and "unplug", perhaps with meditation.

Be Efficient

- Don't set multiple alarms in the morning.
- Go to bed early.
- If you must nap, try doing so in the early afternoon – either a short nap of less than 30 minutes or a recovery nap of around 90 minutes for a full sleep cycle.

MAIN POINTS

- The gut and brain are highly interconnected through the vagus nerve, which sends four times as much information from the gut to the brain.
- There is a strong connection between gut health and cognition, including mood. The implications are that a good balance of gut microbiota is essential for optimum health and good decision-making.
- Practitioners with good resilience in disaster management typically have good social support, a strong sense of their own self-efficacy and physical fitness, and find meaning and purpose in what they do.
- Creating distinct transitions between focus of working in disasters and strategies to rest and recover from such conditions are important for resilience and enhance practitioners' effectiveness and help maintain mental clarity.
- The role of sleep is critical and strategies to improve sleep quality include creating a conducive sleep environment by removing devices, keeping the bedroom cool, and avoiding work in the bedroom.
- A tool is provided to support winding down, which is essential to reduce support embodied well-being.

REFERENCES

Bechara, A., & Damasio, A. R. (2005). The somatic marker hypothesis: A neural theory of economic decision. *Games and Economic Behavior, 52*, 336–372.

Brassey, J., De Smet, A., & Kruyt, M. (2022). *Deliberate calm: How to learn and lead in a volatile world.* HarperCollins.

Dalton-Smith, S. (2017). *Sacred rest: Recover your life, renew your energy, restore your sanity.* FaithWords. Retrieved April 28, 2025, from https://ichoosemybestlife.com/sacred-rest/

Damasio, A. (1999). *The feeling of what happens: Body, emotion and the making of consciousness.* Harcourt Brace.

Dantas, A., Jiao, P., Sack, A., Bruggen, E., & Schuhmann, T. (2021). *A gut feeling: How your gut and brain changes your risk and time preferences.* Maastricht University. https://doi.org/10.21203/rs.3.rs-957480/v1

Dinan, T. G., Stilling, R. M., Stanton, C., & Cryan, J. F. (2015). Collective unconscious: How gut microbes shape human behavior. *Journal of Psychiatric Research, 63*, 1–9.

Falkenstein, M., Simon, M.-C., Mantri, A., Weber, B., Koban, L., & Plassmann, H. (2024). Impact of the gut microbiome composition on social decision-making. *PNAS Nexus, 3*(5). https://doi.org/10.1093/pnasnexus/pgae166

Ferguson, A. (2023). *The Vagus nerve reset.* Griffin Press.

Joas, H. (1996). *The creativity of action.* Polity Press.

Lerner, J. S., Li, Y., Valdesolo, P., & Kassam, K. S. (2015). Emotion and decision making. *Annual Review of Psychology*, *66*, 799–823. https://doi.org/10.1146/annurev-psych-010213-115043

Lucas, G. (2018). Gut thinking: The gut microbiome and mental health beyond the head. *Microbial Ecology in Health and Disease*, *29*(2), 1548250.

O'Mahony, S. M., Clarke, G., Borre, Y. E., Dinan, T. G., & Cryan, J. F. (2015). Serotonin, tryptophan metabolism and the brain-gut-microbiome axis. *Behavioural Brain Research*, *277*, 32–48.

Owen, C. (2024). Learning in the air traffic control tower: Stretching co-presence through interdependent sentience. *Educational Philosophy and Theory*, *56*(5), 496–504.

Shafir, T., Tsachor, R. P., & Welch, K. B. (2016). Emotion regulation through movement: Unique sets of movement characteristics are associated with and enhance basic emotions. *Frontiers in Psychology*, *6*, 2030.

Šimić, G., Tkalčić, M., Vukić, V., Mulc, D., Španić, E., Šagud, M., Olucha-Bordonau, F., Vukšić, M., & Hof, P. R. (2021). Understanding emotions: Origins and roles of the amygdala. *Biomolecules*, *11*(6), 823. https://doi.org/10.3390/biom11060823

van Damn, N., & van der Helm, E. (2016). *The organizational cost of insufficient sleep*. McKinsey Quarterly. Retrieved April 28, 2025, from https://www.mckinsey.com/capabilities/people-and-organizational-performance/our-insights/the-organizational-cost-of-insufficient-sleep

Warner, L. (2025). *5 simple ways to improve gut health*. Harvard Health Publishing, Retrieved April 28, 2025, from https://www.health.harvard.edu/staying-healthy/5-simple-ways-to-improve-gut-health

5 Emotional Intelligence and Self-awareness

INTRODUCTION

Evidence of the influence of emotional intelligence (EI) on the quality of decision-making has been increasing (Beling & Wild, 2024; Dilawar et al., 2021; Hess & Bacigalupo, 2011; Magny & Todak, 2021; Syeda et al., 2019). EI refers to the regulation, perception, and management of one's own and others' emotions. Research conducted by Dilawar et al. (2021) showed that emergency services personnel high in EI are better able to manage the stresses associated with their occupations. Moreover, they cite research where those high in EI were able to self-manage and so were better able to call on rational decision-making styles. In addition, because they were emotionally grounded, they were also able to use their intuition more efficiently and effectively to improve the quality of their decisions. Others have demonstrated the positive impacts that leaders with high EI have had on team performance (Michinov & Michinov, 2022). Uy et al. (2023), for example, found a strong association between EI and successful crisis leadership, mitigating emotional stress and fostering cohesion in teams. In short, cultivating high EI for decision-makers under pressure is an asset.

EMOTIONAL INTELLIGENCE

EI, popularized by Daniel Goleman (1995), was initially defined and then later refined by Mayer and Salovey (1997, p. 31) as being

> the ability to perceive accurately, appraise, and express emotions; the ability to access and/or generate feelings when they facilitate thought; the ability to understand emotion and emotional knowledge; and the ability to regulate emotions to promote emotional and intellectual growth.

Daniel Goleman further developed a model of EI, which included four domains, summarized in Figure 5.1.

The four domains are characterized as:

1. Self-awareness
   Self-awareness entails the ability to recognize one's emotions and understand how they impact performance, making it essential for pinpointing both strengths and weaknesses. Practitioners who excel at self-assessment are mindful of their limitations, actively pursue feedback, and learn from their errors. This practice significantly improves their capacity to regulate emotional responses in different contexts (Ismail et al., 2024). In a

DOI: 10.1201/9781003566007-6

Self-awareness	Self-management	Social Awareness	Relationship Management
	Emotional self-control		Influence
		Empathy	
			Coach and mentor
	Adaptability		
Emotional self-awareness			Conflict management
	Achievement orientation		
		Organizational awareness	Teamwork
	Positive outlook		Inspirational leadership

FIGURE 5.1 Emotional intelligence domains and competencies. (From Goleman, D., & Boyatzis, R. E. (2017). Emotional intelligence has 12 elements. Which do you need to work on? *Harvard Business Review*. Retrieved April 28, 2025, from https://hbr.org/2017/02/emotional-intelligence-has-12-elements-which-do-you-need-to-work-on)

nutshell – knowing what one is feeling and why provides a good basis for decision-making and provides us with the foundations of a moral compass.

2. Self-management

   Self-management is centered on regulating internal states, impulses, and personal resources. Key competencies within this framework include

   - emotional self-control, which aids in maintaining composure during stressful circumstances;
   - trustworthiness, which reflects a consistent alignment with personal values;
   - conscientiousness, which pertains to responsible management of duties;
   - adaptability, which involves implementing creative solutions;
   - achievement orientation, which fuels motivation for success; and
   - initiative, which focuses on proactively preventing problems (Ismail et al., 2024).

   In short, when we have good self-management, we can handle distressing emotions in effective ways, so they don't cripple us. Since every emotion has a cognitive function (see Chapter 4), we can attune to them and thus learn from them. Self-management is also about marshalling positive emotions. That is, getting enthused and aligning actions with our passions.

3. Social awareness

   Social-emotional awareness encompasses various competencies, including empathy, which fosters a deeper understanding of the needs and emotions of others; service orientation, which ensures that customer needs are effectively aligned with available services; and organizational awareness, which enables practitioners to interpret situations objectively, free from bias. These competencies are essential for building strong interpersonal relationships and effectively navigating social dynamics. In short, it is about knowing what others are feeling and provides the basis for political astuteness (see Chapter 18).

4. Relationship management
   Relationship management encompasses a set of skills designed to elicit positive responses from others. Key competencies in this area include the ability to influence, communicate effectively, manage conflicts, demonstrate visionary leadership, act as a catalyst for change, build strong connections, and promote collaboration and teamwork (Oliveira, 2024). Mastery of these skills is essential for fostering successful interpersonal interactions and effective leadership within any organization.

   Social awareness and relationship management are about understanding and influencing others effectively. In the chapter on influence (see Chapter 11), we will also examine the ways in which influence is about understanding not just the formal organizational structure but also the informal leaders and power brokers who need to be on side to get things done.

Linkages Between the Elements of Emotional Intelligence

The relationship between key elements is highlighted in Figure 5.2.

According to research (Boyatzis et al., 2017), what team members look for in leaders is inspirational leadership – which is akin to providing a goal members want to work toward, as well as empathy and compassion (that is, the leader has their interests at heart). Members also desire to work with leaders who display emotional self-control, are open to adaptability, and express humanizing thinking about others

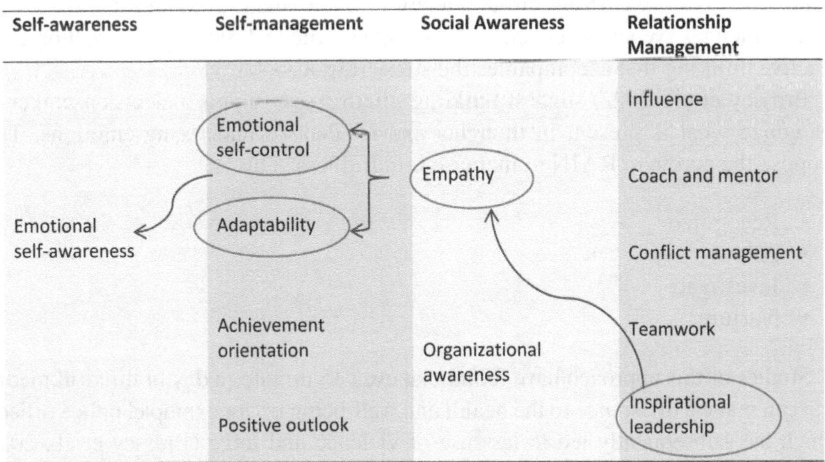

FIGURE 5.2 The relationship between key elements. (From Goleman, D., & Boyatzis, R. E. (2017). Emotional intelligence has 12 elements. Which do you need to work on? *Harvard Business Review*. Retrieved April 28, 2025, from https://hbr.org/2017/02/emotional-intelligence-has-12-elements-which-do-you-need-to-work-on)

(that is, use humanizing language such as kids, family, community rather than dehumanizing language, i.e., turning people into categories like for example, "our personnel are key resources" (Boyatzis et al., 2017, p. 509). The key thing here, however, is that none of these is possible without decision-makers being grounded and aware of themselves (see the directional arrows added in Figure 5.2).

The rest of this chapter will examine how decision-makers under pressure can develop emotional self-regulation and increase emotional self-awareness. This is an important skill particularly in high-pressure environments, not only to navigate VUCA but also because of the negative consequences of mismanagement.

UNDERSTANDING EMOTIONAL SELF-AWARENESS

In a similar way to that suggested in Chapter 2, by Lisa Frye – that we can become mindfully metacognitive about where our attention is focused – Brassey et al. (2022) suggest that when we are confronted with difficult situations, we can react in one of two ways. We can engage in either a "protective" mode or an "adaptive" mode. Under these circumstances, it is easy to hunker down and contract into a "protective mode" using old established patterns of being and thinking. We are more likely to want to protect our identities, opinions, stories, and inner logic. We use old patterns to solve problems. We defend and cling tightly to these instead of considering the possibility of trying or learning something new. In an adaptive mode, we can break free of conventional problem-solving methods and innovate.

There are things we can do when faced with difficult circumstances where the default position is to move into a "protective mode" where an adaptive mode is going to be more productive. They are, in part, about becoming more mindful of the drive to be protective and to pause prior to reacting. Through this cycle, we can pivot from protective mode to learning through adapting even when faced with the most complex challenges. When practiced with awareness, the techniques will ameliorate the reactive thinking that accompanies the stress response.

Brassey et al. (2022) suggest building mindfulness supports decision-makers to recognize what is present in thoughts and to identify underlying emotions. They propose the acronym RAIN to increase mindfulness. This is:

- Recognize
- Allow
- Investigate
- Nurture

Studies of this approach have found that even 15 minutes a day of mindful meditation can make a difference to the health and well-being of, for example, police officers, which has subsequently led to less use of violence and force (Brassey et al., 2022). This is because such approaches assist in cultivating better thinking and thus decision-making. Stopping and engaging in these mindfulness behaviors enables us to tap into what is happening in a difficult situation and to become emotionally self-aware rather than shutting down when responding to emotional triggers. Tool 5.1 provides some suggestions about what to look for on reflection to self-assess your own triggers.

TOOL 5.1 SELF-ASSESSMENT OF EMOTIONAL TRIGGERS

VIEW CHALLENGES AS OPPORTUNITIES

In protective mode, we tend to frame things negatively and see challenges or surprises as problems. Reframing can open up a whole new set of possibilities.

SEEK FIRST TO UNDERSTAND

Pausing what is going on helps us to stay open so we can try to understand what is going on for other people. The "reality" that applies to us does not always apply to others. They have their own contexts and hidden icebergs causing them to act in certain ways. Simply asking for more context about what is going on for others helps build a shared understanding. This can open us up to other realities, not just our own.

GET CURIOUS

Asking questions to ourselves or others causes us to slow down, reflect, and engage the executive functioning parts of our brains. It also keeps us from reacting in the same ways as always and forces us to see new possibilities. Instead of being an expert who must have all the answers, reframe to be a curious learner who must ask the right questions.

Source: Brassey et al. (2022).

Staying Emotionally Present

Staying emotionally present with others who have been affected by a disaster or crisis can be hard. This is particularly the case for decision-makers who want to "get things done".

Practitioners under pressure need to acknowledge the challenges that others face without inadvertently making things worse by attempting to take on and remove the distress of others. This can sometimes happen if the empathy leads a decision-maker into taking up a "heroic savior" mindset, which is often the last thing affected people need. This can happen sometimes when decision-makers become overwhelmed with the pain and suffering of others. When this happens, they try to step in and fix the issues for those affected, taking on a "rescuer/savior" mindset in order to take the pain away from those affected. This understandable "default" position, while laudable, can do more harm than good. This means decision-makers need to think about how the disaster or crisis they are managing will personally influence their own actions – the motive to "do something and fix" other people's pain and suffering, while an understandable motive is coming from the wrong intention. It can also make things worse because it establishes unrealistic expectations and disempowers those impacted. In the research conducted by Moreton (2018, p. 19), one community participant described the experience in the following way.

> We needed help, but we weren't helpless. We needed someone to come along and hold our hands, with the tools and support that we needed, but knowing when to take their hands away. We didn't want people to come in and take over. Part of going through the process was to feel that we had some strength.

Becoming humble and realizing "this is not about me" will assist decision-maker leaders under pressure to sit beside others and to learn from those affected. Chapter 16 further explores the dynamics of those individuals and communities impacted by psycho-social trauma. It is also important to be mindful that when operating under pressure practitioners often require a good dose of self-compassion.

Practicing Self-compassion

Stephen Thompson, a Director at the Headington Institute, has written several short articles to assist those responding to the impacts of the LA fires of 2024/2025. In one such article, titled "Self-compassion for wildfire responders: Managing guilt, compassion fatigue and responsibility" (Thompson, 2025), he notes the following strategies which we believe support remaining emotionally self-aware.

Maintaining Self-compassion

It is important to treat oneself with the same kindness and respect you would offer a friend or colleague. This is not self-pity or self-indulgence, but rather, is acknowledging pain and offering yourself what Thompson (2025) describes as grace instead of trying to "tough it out".

Managing Guilt

Decision-makers under pressure can often feel guilty for having failed others, even when the outcomes were not within one's control. This often stems from having unrealistically high expectations, rather than acceptance of human limitations.

Recognizing Compassion Fatigue

When exposed to human suffering, sometimes for days and weeks, it is important to recognize and acknowledge compassion fatigue. Compassion fatigue is evident in the following symptoms:

- Emotional numbness or irritability
- Difficulty sleeping or concentrating
- Feeling detached from your work or others
- Increased use of substances to cope

Recognizing these signs early is a key step toward recovery.

Guarding against Responsibility Overload

It is often the case that decision-makers under pressure carry the weight of their responsibilities and also the responsibilities of those reliant on them. This includes survivors, colleagues, and family members. While noble, it can become overwhelming. It is important to care for oneself as you cannot sustain the mission without sustaining yourself.

Included in Tool 5.2 are the strategies for practicing self-compassion identified by Thompson (2025).

TOOL 5.2 STRATEGIES FOR PRACTICING SELF-COMPASSION

Element	Practice
Notice without judgment	When you encounter challenging thoughts or emotions, such as guilt or helplessness, take a moment to acknowledge them without criticism. Simply label what you're experiencing without an urge to fix or dismiss it. You might say to yourself: "This is tough right now, and I'm feeling [insert emotion]. It is okay to feel this".
Shift the inner voice	If you find yourself thinking things like "I should have done more", try to reframe those thoughts. Imagine what you would say to a teammate in a similar situation and extend that same kindness to yourself.
Use micro-moments of reset	Even short intervals, such as 60 seconds of deep breathing, stretching, or grounding techniques, can make a difference (see also Chapter 4). You don't need elaborate self-care routines to reset your mindset. For example, practice this breathing technique: Inhale for 4 seconds, hold for 4, and exhale for 6. Repeat this three times, allowing it to suffice for the moment.
Reconnect with purpose, not perfection	Focus on the underlying reasons for your work rather than the results that are beyond your control. Remember that your presence and efforts matter more than you might realize.
Create a circle of support	Reach out to someone who understands what you're going through. Engaging in peer support, even for a brief five-minute conversation, can help you process your emotions and reinforce that you are not alone in your experiences.

Source: Thompson (2025).

MAIN POINTS

- EI plays a pivotal role in enhancing decision-making quality, particularly under high-stress conditions.
- Research underscores the benefits of high EI in managing occupational stresses, facilitating rational decision-making, and effectively utilizing intuition. The four domains of EI, as outlined by Goleman (1995) – Self-Awareness, Self-Management, Social Awareness, and Relationship Management – are crucial for optimal decision-making.
- Cultivating high EI is an important skill for decision-makers operating under pressure, leading to improved outcomes and enhanced team performance.
- Practicing self-compassion enables the decision-maker to remain emotionally grounded and present.

REFERENCES

Beling, C., & Wild, G. (2024). The association between emotional intelligence and decision making for pilots. *Journal of Air Transport Management, 114*, 102506.

Boyatzis, R. E., Thiel, K., Rochford, K., & Black, A. (2017). Emotional and social intelligence competencies of incident team commanders fighting wildfires. *The Journal of Applied Behavioral Science, 53*(4), 498–516.

Brassey, J., De Smet, A., & Kruyt, M. (2022). *Deliberate calm: How to learn and lead in a volatile world*. HarperCollins.

Dilawar, S. M., Durrani, D. K., Li, X., & Anjum, M. A. (2021). Decision-making in highly stressful emergencies: The interactive effects of trait emotional intelligence. *Current Psychology, 40*(6), 2988–3005. https://doi.org/10.1007/s12144-019-00231-y

Goleman, D. (1995). *Emotional intelligence: Why it can matter more than IQ*. Bloomsbury Publishing.

Goleman, D., & Boyatzis, R. E. (2017). Emotional intelligence has 12 elements. Which do you need to work on? *Harvard Business Review*. Retrieved April 28, 2025, from https://hbr.org/2017/02/emotional-intelligence-has-12-elements-which-do-you-need-to-work-on

Hess, J. D., & Bacigalupo, A. C. (2011). Enhancing decisions and decision-making processes through the application of emotional intelligence skills. *Management Decision, 49*(5), 710–721. https://doi.org/10.1108/00251741111130805

Ismail, F., Mohamad Azmi, N. A. H., & Chan, S. W. (2024). Emotional intelligence and psychological well-being among police officers. *Advances in Humanities and Contemporary Studies, 5*(3), 34–41.

Magny, O., & Todak, N. (2021). Emotional intelligence in policing: A state-of-the-art review. *Policing: An International Journal, 44*(6), 957–969.

Mayer, J. D., & Salovey, P. (1997). What is emotional intelligence? In P. Salovey & D. J. Sluyter (Eds.), *Emotional development and emotional intelligence: Educational implications* (pp. 3–34). Basic Books

Michinov, E., & Michinov, N. (2022). When emotional intelligence predicts team performance: Further validation of the short version of the workgroup emotional intelligence profile. *Current Psychology, 41*(3), 1323–1336.

Moreton, M. (2018). 'We needed help, but we weren't helpless': The community experience of community recovery after natural disaster in Australia. *The Australian Journal of Emergency Management, 33*(1), 19–22.

Oliveira, J. (2024). Building emotional intelligence: Strategies for effective communication and relationship management. *Empowerverse*. https://doi.org/10.5281/zenodo.10888838

Syeda, M. D., Durrani, D. K., Li, X., & Anjum, M. A. (2019). Decision-making in highly stressful emergencies: The interactive effects of trait emotional intelligence. *Current Psychology, 40*, 1–18. https://doi.org/10.1007/s12144-019-00231-y

Thompson, S. (2025). *Self-compassion for wildfire responders: Managing guilt, compassion fatigue, and responsibility*. Headington Institute. Retrieved April 28, 2025, from https://headington-institute.org/blog/resource/self-compassion-for-wildfire-responders-managing-guilt-compassion-fatigue-and-responsibility/

Uy, F., Kilag, O. K., Abendan, C. F., Macapobre, K., Cañizares, M. C., & Yray, F. (2023). Application of adaptive crisis management theory: The dynamics of leadership in times of crisis. *Excellencia: International Multi-disciplinary Journal of Education (2994-9521), 1*(5), 159–170.

6 Decision-making and Ethics

INTRODUCTION

In significant crises, the very core of our value systems is sometimes questioned or even under threat. This can occur for individuals, groups, communities, or whole societies. These kinds of crises include, for example, the experience of living through the COVID-19 pandemic and past wars and conflicts.

Ethics are at the heart of the way practitioners translate their values into everyday practice. They guide both our personal and professional lives. Although personnel working in VUCA situations such as emergency management are rated highly in the community for trustworthiness, there has been considerable scrutiny of leadership and decision-making, and some have observed a recent erosion of public trust (Bell et al., 2025).

Acting ethically involves acting with integrity, fairness, respect, and honesty. Sometimes decision-making under pressure leads to moments where dilemmas occur, and decision-makers are reduced to choosing between the least-worst options (Santos et al., 2021). It is important to examine some of the ways we make decisions to identify how they are always underpinned by values – whether we are conscious of them or not. Sometimes even well-intentioned leaders can make poor decisions based on several red flag conditions.

TRIGGERS DERAILING ETHICAL DECISION-MAKING

Sometimes decision-makers under pressure can get derailed by several factors. These can be as simple as the decision-maker relying too often on a leadership attribute that has become an over-used strength. This becomes a "default" position, and the decision-maker, operating almost automatically, can miss opportunities that may build capability in others to create even better outcomes. In other words, enormously important decisions made by intelligent, responsible people with the best information and intentions are sometimes hopelessly flawed. This is in part due to the way our brains work through pattern recognition and emotional tagging as previously discussed in Chapters 2 and 3. Both these behaviors are normally reliable however, in certain circumstances, either one can skew judgment.

PATTERN RECOGNITION

Faced with a new situation, we make assumptions based on prior experiences and judgments (Finkelstein et al., 2009). Thus, a chess master can assess a chess game and choose a high-quality move in as little as six seconds by drawing on patterns they have seen before. But pattern recognition can also mislead us. When we're

DOI: 10.1201/9781003566007-7

dealing with seemingly familiar situations, our brains can cause us to think we understand them when we don't.

Emotional Tagging

This is the process by which emotional information attaches itself to the thoughts and experiences stored in our memories (see also Chapter 5). This emotional information tells us whether to pay attention to something or not, and it tells us what sort of action we should be contemplating (immediate or postponed, fight or flight). We can leap to conclusions and are reluctant to consider alternatives; we are particularly bad at revisiting our initial assessment of a situation (Finkelstein et al., 2009).

STRATEGIES FOR PRACTITIONERS

In analyzing how well-intentioned leaders can make bad judgments, Finkelstein et al. (2009) found that they were affected in all cases by three factors that either distorted their emotional tags or encouraged them to see a false pattern, factors called "red flag conditions" (Finkelstein et al., 2009). It is important that practitioners have awareness about what kinds of conditions might derail their thinking and their decisions and that they create the conditions for team members to be able to raise the red flag. Understanding why even well-intentioned leaders can stray off course in terms of their moral compass is important. Tool 6.1 highlights some of the common red flags that can derail decision-making effectiveness and some reflective questions to consider.

TOOL 6.1 RED FLAGS AND QUESTIONS TO SELF-ASSESS TRIGGERS

The presence of inappropriate self-interest

Typically, biases are caused from the emotional importance we place on information, which in turn makes us more ready to perceive the patterns we want to see. Research has shown that even well-intentioned professionals, such as doctors and auditors, are unable to prevent self-interest from biasing their judgments of which medicine to prescribe or opinion to give during an audit.	Question: To what degree do I have an interest in the problem and the potential outcome?

The presence of distorting attachments

We can become attached to people, places, and things, and these bonds can affect the judgments we form about both the situation we face and the appropriate actions to take.	Question: Do I have a default to prioritize something in this situation based on my past?

The presence of misleading memories

These are memories that seem relevant and comparable to the current situation but lead our thinking down the wrong path. They can cause us to overlook or undervalue some important differentiating factors. The chance of being misled by memories is intensified by any emotional tags we have attached to the past experience. If our decisions in the previous similar experience worked well, we'll be more likely to overlook key differences.	Question: Is this issue triggering something from my own previous experience?

Source: Finkelstein et al. (2009).

Finkelstein et al. (2009) discuss how practitioners can overcome red flags if they remain mindful of their own needs, wants, desires, and biases and consider the needs of others "with grace, compassion and empathy" (p. 178).

Decision-making and Moral Injury

Moral injury occurs when individuals experience, witness, or fail to prevent events that violate deeply held moral beliefs. Practitioners working in high-consequence VUCA-related conditions are routinely exposed to human suffering and need to make quick, morally challenging decisions that can affect their well-being (Thomas & Bizumic, 2025). According to Nash et al. (2013), moral injury can relate to

- Perceived transgressions – that is, acts of commission or omission that violate personal morals or values
- Perceived betrayals – that is, feeling let down by leaders, peers, or institutions once trusted

According to Lentz et al. (2021), participating in or observing an event or situation that conflicts with personal values can potentially lead to the development of moral injury. It is important, therefore, that practitioners can draw on ethical principles wherever possible to guide them in their decision-making. Tool 6.2 discussed provides a guiding framework.

Making Decisions Using Ethical Principles

Sakakushev (2021) makes the point that every high-consequence strategy is dependent on ethical attitudes and approaches, which arise from complex dilemmas, requiring definite codes of conduct. It is therefore important to have some knowledge and self-awareness about how motivation to act in certain ways and reactions can be based on ethical principles. Sakakushev (2021) observed that during a crisis ethical practice should be based on humanitarian assistance, information and opportunities to participate, respect of dignity and persons, emergency assistance for the most vulnerable persons, measures to safeguard and rehabilitate the environment, strengthening resilience to the effects of disasters, and protection of economic, social, cultural, civil, and political rights.

Etkin and Timmerman (2013, p. 277) contend that persons affected by disasters can sometimes be "treated as objects without rights and to whom duties are not owed, instead of subjects with rights and to whom duties are owed". They propose that this is increasingly important in a world where the assumption of a return to a pre-existing normalcy may become increasingly invalid in contemporary social systems. Under these circumstances, it is critically important that decision-makers are aware of the ethical impacts of their decisions and the consequences of those decisions.

Tool 6.2 (adapted from Etkin, 2021) provides some reflective questions when engaged in decision-making involving ethical dilemmas.

In VUCA environments, practitioners require what Tomkins (2020, as cited in Wilson, 2023, p. 176) refers to as an "ethics of care which fosters an empathetic attunement to people's emotional needs". This requires practitioners to be sensitive to different hopes, fears, and a sensitivity to different world views and modus operandi. These elements are often referred to as relational leadership. As discussed in Chapter 5, sometimes practitioners can undermine the support affected people need by a misguided sense of trying to be the hero or savior who can somehow take the pain away. One strategy that supports practitioners to have empathetic attunement without overreaching is to engage in the practice of mindful listening.

TOOL 6.2 DECISION-MAKING PROCESS FOR ETHICAL DILEMMAS

Context:

- Know the facts of the case and the antecedents
- Know the expectations of those involved
- Be aware of the social and political pressures on decision-makers and stakeholders

Nature of the decision-maker:

- What are the biases and values of the decision-maker?
- What knowledge do they have of ethics and relevant codes?
- What relevant experience do they have?
- What is their character like?

Choice of actions and implementations:

- Consider which action, choice, or combination of choices is best defended by ethical arguments
- Identify factors that might impede a chosen ethical action
- Consider how these impediments might be overcome
- Eliminate impediments where practical and implement ethical actions

Problem clarification:

- Identify relevant rights, duties, responsibilities, and cultural norms/values
- Identify pressures on decision-makers and stakeholders
- Identify other relevant information needed to clarify the situation and possible actions

Recognition of the problem as ethical:

- To what extent does the problem evoke ethical concerns?
- To what extent does the problem evoke personal and cultural values?
- Apply the following three intuitive tests:
  - The pillow test – would your decision allow you to sleep at night?
  - The child test – what would you tell your child to do?
  - The media test – would you want your decision to be in the public?

Action identification and evaluation:

- Generate action possibilities:
  - Determine the consequences of the various actions
  - Examine the rights, duties, norms and values associated with the actions
  - Develop an initial justification for each possibility
- Assess the justifications:
  - Is the approach being used in a comprehensive and logical way?
  - Is the argument consistent?
  - Are all ethical concerns identified?
  - Is there moral self-awareness?

Source: Etkin (2021).

MINDFUL LISTENING

One strategy helpful for practitioners is to become more mindful of how we are attuning to the needs of others in order to act morally and ethically. According to Scharmer (2016), listening is one of the most underappreciated leadership skills, often leading to a disconnect between the ethical decision-maker and their situations. Scharmer outlines four distinct levels of listening:

Level 1: Listening from Habit (Downloading.) This level involves listening only for what you already know, effectively reconfirming your existing opinions and judgments. It stems from the assumption that you already know and understand what is being said, causing you to rely on familiar patterns and mental models. The attention is really on your own story.

Level 2: Factual Listening. At this level, you listen with a somewhat open mind, focusing on picking up new information and noticing differences. This allows you to identify disconfirming data – things that are new or deviate from what you already know. Looking for facts means if the facts are not available, we can be dismissive. Your listening comes from your own worldview.

Level 3: Empathic Listening. Empathic listening involves opening your heart to fully understand another person's perspective. It requires setting aside your own agenda to see the situation through their eyes. This type of listening builds a deeper connection as you experience what it is like to "stand in someone else's shoes". Listening here comes from the world of the speaker.

Level 4: Generative Listening. This highest level of listening involves connecting to emerging possibilities and requires an open will. It fosters a shift in identity and from self, leading to connectedness with others. We are picking up on what is said and what is not said, being open to anything emerging. In group dynamics, this is described as synergy, while in interpersonal communication, it manifests as a deep sense of connection and creativity.

Enemies of Listening

According to Scharmer (2016), there are three key obstacles to effective listening – referred to as the "enemies of listening". These are internal voices that hinder deeper engagement with the situation at hand:

1. *Voice of Judgment*: This voice closes your mind by dismissing new ideas and information, often to reinforce your existing worldview and mental models.
2. *Voice of Cynicism:* This voice shuts down your heart by fostering skepticism about others' intentions or outcomes. It stems from the belief that trusting only yourself is the safest option.
3. *Voice of Fear:* This voice inhibits openness by clinging to the past or present, stifling your ability to let go and take risks. It thrives on the fears that hold you back and prevent progress.

Overcoming the "enemies of listening" requires intentional effort and self-awareness. Some strategies to address each one include:

1. *Voice of Judgment:*
   - Practice active listening by focusing fully on the speaker without forming immediate opinions.
   - Approach conversations with curiosity, asking open-ended questions to understand different perspectives.
   - Reflect on your own biases and assumptions, challenging them when they arise.
2. *Voice of Cynicism:*
   - Cultivate empathy by putting yourself in others' shoes and considering their intentions.
   - Focus on finding common ground or shared goals in conversations.
   - Remember that trust builds over time through dialogue and collaboration.
3. *Voice of Fear:*
   - Embrace vulnerability by acknowledging your fears and consciously letting go of the need for control.
   - Encourage a growth mindset, viewing challenges and risks as opportunities for learning.
   - Take small steps toward openness, celebrating progress along the way.

Practicing mindfulness can also help quiet these inner voices, allowing you to stay present and engaged during conversations, leading to a greater capacity to act with ethics and integrity.

ETHICS AND APPLICATION OF ARTIFICIAL INTELLIGENCE

There is no doubt that the application of artificial intelligence (AI) into an increasing number of domains, and in particular in emergency management preparedness, response, and recovery, is transforming practice and offers unprecedented capabilities. For example, AI can support prediction by analyzing vast sets of data to predict possible catastrophes, support timely evacuations through automating communications and warnings, and enhance situation awareness and resource allocation.

However, with these opportunities come significant ethical challenges. Visave (2025) reports on a survey of emergency management AI systems in the US that highlight how 68% lacked adequate documentation of their data sources, while 42% failed to provide clear justifications for their recommendations. He concludes that the use of AI as a "black box" has serious implications, particularly "when flood prediction models disproportionately affect vulnerable populations or when opaque decision-making processes lead to suboptimal resource allocation during critical rescue operations" (Visave, 2025, p. 3967). Ultimately, AI systems are trained on datasets that reflect existing societal inequalities and biases, which can lead to biased decision-making and unequal treatment (Visave, 2024).

As leaders, practitioners relying on decision-making using AI in VUCA-related events have an ethical responsibility to ensure such systems are transparent and explainable to stakeholders in how decisions have been reached. They also have a duty of care to ensure that information used in AI decision-making is protected from the misuse of personal information. Here, the balance is between transparency and data security. Visave (2025) contends that any use of AI needs to be explainable where the algorithms used to reach conclusions can be evaluated. He proposes the following guidelines to support a transparency framework for the use of AI in decision-making. When such systems are being developed, it is incumbent on organizations to ensure that the following can be explained to all stakeholders.

1. **Accountability:** Decision-making processes can be clearly defined and explained to all stakeholders.
2. **Tailored Communication:** These explanations are developed for both technical and non-technical audiences.
3. **Decision Standards:** The underlying rationale for any criteria used to drive decisions can be clearly communicated
4. **Risk Management:** The potential system risks, biases, and mitigation strategies are openly disclosed.

Such systems must be open to audit of their performance and regularly reported in terms of performance-based metrics. Challenges remain however for how decision-makers can maintain real-time transparency for scrutiny in time-critical situations.

MAIN POINTS

- The chapter emphasizes the critical role of moral values and ethics in leadership, especially during crises.
- Crises challenge foundational value systems, requiring practitioners to navigate complex ethical dilemmas with integrity, fairness, and empathy.
- Practitioners are at risk of moral injury if they are faced with decisions that violate their own moral beliefs.
- The concept of an "ethics of care" focuses on understanding and addressing the emotional needs of individuals affected by crises.
- Being aware of potential "red flag conditions" (inappropriate self-interest, distorting attachments, and misleading memories) supports more informed and ethical decisions.

- Mindful listening is highlighted as an essential leadership skill that fosters meaningful connections with teams and stakeholders.
- Decisions should be informed by a deep understanding of their ethical implications and the diverse perspectives of those impacted.
- This includes the application of AI in decision-making. AI systems must be explainable and their algorithms open to scrutiny for ethically accountable decision-making in VUCA situations.

REFERENCES

Bell, V., Murray, E., Muñoz, L., & Krahé, C. (2025). In harm's way: Moral injury and the erosion of trust for emergency responders in the United Kingdom. *European Journal of Psychotraumatology, 16*(1), 2513107.

Etkin, D. (2021). *The ethical emergency manager: Issues, morality and dilemmas* (2nd ed.). Amazon.

Etkin, D., & Timmerman, P. (2013). Emergency management and ethics. *International Journal of Emergency Management, 9*(4), 277–297.

Finkelstein, S., Whitehead, J., & Campbell, A. (2009). *Think again: Why good leaders make bad decisions and how to keep it from happening to you.* Harvard Business Press.

Lentz, L. M., Smith-MacDonald, L., Malloy, D., Carleton, R. N., & Brémault-Phillips, S. (2021). Compromised conscience: A scoping review of moral injury among firefighters, paramedics, and police officers. *Frontiers in Psychology, 12*, 639781.

Nash, W. P., Carper, T. L. M., Mills, M. A., Au, T., Goldsmith, A., & Litz, B. T. (2013). Psychometric evaluation of the moral injury events scale. *Military Medicine, 178*(6), 646–652.

Sakakushev, B. E. (2021). Current ethical dilemmas and code of conduct in disasters. In E. Pikoulis, & J. Doucet (Eds.) *Emergency medicine, trauma and disaster management. Hot topics in acute care surgery and trauma.* Springer. https://doi.org/10.1007/978-3-030-34116-9_45

Santos, L. S., Lebioda, L., & Serafim, M. C. (2021). How to deliberate when everything is chaos? The moral dilemmas and ethical decision-making in public management in emergencies. *Revista de Contabilidade e Organizações, 15*, 1–12. www.redalyc.org/articulo.oa?id=235267639006

Scharmer, O. (2016). Theory U: *Leading from the future as it emerges. The social technology of presencing.* Berrett-Koehler Publishers.

Thomas, V., & Bizumic, B. (2025). Moral injury in health workers, emergency services, police, government officials, and teachers: Measurement invariance of the occupational moral injury scale (OMIS) and group comparisons. *Psychological Trauma: Theory, Research, Practice, and Policy.*

Tomkins, L. (2020). Paradox and power in caring leadership: Critical and philosophical reflections. Edward Elgar Publishing. In Wilson, S. (2023). Leadership in a VUCA context: Some foundational considerations. *Journal of Applied Journalism & Media Studies, 12*(2), 169–183. https://doi.org/10.1386/ajms_00112_1

Visave, J. (2024). AI in emergency management: Ethical considerations and challenges. *Journal of Emergency Management and Disaster Communications, 5*(1), 165–183.

Visave, J. (2025). Transparency in AI for emergency management: Building trust and accountability. *AI and Ethics, 5*, 3967–3980.

7 Leadership in VUCA

INTRODUCTION

When working under pressure, the skills of leading through deep uncertainty are vitally important. As the previous chapters have demonstrated, decision-makers leading in volatile, uncertain, complex, and ambiguous (VUCA) conditions confront demands far more complex than those historically faced by their predecessors (Marcus et al., 2014). According to Black et al. (2024), changes in complexity will also require changes in styles of leadership to cultivate decision-makers who are more comfortable with navigating complex adaptive systems.

Heifetz (1994), Heifetz and Linsky (2017), and Heifetz et al. (2009) argue that leaders in VUCA-related contexts need to develop a set of tools, ideas, and frameworks to be able to analyze and intervene within social systems to help make progress on tough, challenges. Leaders play an important role in enabling good outcomes, particularly in crisis contexts. Understanding oneself is one of the most critical skills a leader can have. This includes understanding the subjective frames we bring to a situation and critically evaluating their strengths and weaknesses. It also includes recognizing the impact we have on others (see Chapter 5).

Under these conditions, leaders need to navigate several tensions. Jordan et al. (2020) describe these as tensions between:

- Being an expert and being a learner.
- Being constant and being an adaptor.
- Being directive and being a listener.
- Holding the power and sharing the power.
- Trusting intuition and being analytical.
- Striving for perfection and being an accelerator.

The qualities leaders need to balance these tensions and act accordingly include self-awareness, a focus on learning and adaptation, and understanding the context (Jordan et al., 2020). Being able to embrace these paradoxes also requires intellectual humility and political astuteness (see Chapter 13).

INTELLECTUAL HUMILITY

> True merit is like a river. The deeper it is, the less noise it makes.
> *Edward Frederick Halifax*

According to Black et al. (2024), being humble is a necessary skill to tackle complex problems. Intellectual humility is present when leaders recognize they do not know all and do not have all the answers. Admitting this and seeking

DOI: 10.1201/9781003566007-8

others' input requires some humility. While still quietly confident, intellectually humble leaders avoid arrogant pride and ego.

Intellectually humble leaders can acknowledge the limitations of data, tools, and thinking capacity, recognizing the lack of experience in addressing novel issues. Another mark of a leader who practices humility is the leader's treatment of others. Such leaders treat everyone with respect regardless of position. In using the Boston Bombings as a case example, Starr (2020) discussed the way in which the Governor who assembled the Operations Centre ignored his boss's advice to *let everyone know you're the most senior officer* and instead went around the room asking each person *how can I help*?. This example set the tone for a unity of purpose and enabled a spirit of generosity where each participant seeks others to succeed (later identified by Marcus et al., 2014, as a key contributor to "swarm intelligence").

Intellectually humble leaders will also be comfortable with others challenging their ideas, encouraging them as needed. According to Black et al. (2024), this involves avoiding surprise by proactively interrogating assumptions and biases and co-adaptation, which means iterating as needed. This means practitioners need to be aware of the challenges of leading in deep uncertainty and the risks that can be engaged with pride and ego.

CHALLENGES OF LEADING IN DEEP UNCERTAINTY

Leaders managing uncertainty and under pressure face a number of challenges. Stern et al. (2020) observe that while arrangements help identify roles and responsibilities in a crisis, they do not provide guidance on how to manage and navigate the crucial interpersonal and inter-professional processes that support – or inhibit – problem-solving in VUCA environments. The following four challenges are a synopsis of Stern et al.'s work for the Australian and New Zealand School of Government's *Leading in a Crisis* series. The challenges leaders need to develop awareness about include:

Challenge 1: Overbearing Leadership

In high-pressure situations, leaders often feel the urge to take a heroic, command-and-control approach (see also Owen et al., 2015, 2025). While this method may appear effective and expedient on the surface, Stern et al. (2020) argue that it actually hinders performance. Leaders who are overbearing tend to dominate discussions, limit the flow of information, promote conformity among team members, and undermine the authority, expertise, and critical thinking skills of team members and other stakeholders. To address this challenge, Stern et al. offer several recommendations for leaders:

- **Adapt Deliberation and Decision-Making Processes:** Leaders should tailor their processes to fit the available time frames, fostering purposeful dialogue and encouraging inclusive participation from all team members.
- **Diversify Team Composition:** It is important to ensure a mix of professions, expertise, and backgrounds at the decision-making table. Leaders should regularly reassess and adjust team composition based on the evolving nature and demands of the current crisis.

- **Refrain from Dictating Outcomes:** Instead of getting trapped in the role of "the decider", leaders should provide overarching guidance on the group's mission and objectives without revealing personal policy preferences or anticipated decisions too early in the process.
- **Empower Team Members:** Each team member must feel encouraged to contribute authentically, which includes voicing concerns, expressing dissent, and knowing that their perspectives will be heard and valued.

By implementing these practices, practitioners can create a more collaborative and effective team environment, ultimately enhancing performance and decision-making during challenging times.

Challenge 2: Blind Spots and Biases

Stern et al. (2020) note that many crises have been mismanaged in good faith by top teams making fatal judgment calls based on false premises and misperceptions, rather than by willful pursuit of questionable actions. Crises create challenging information environments, where the pressure to make quick decisions can lead to errors. This urgency often encourages instinctive reactions instead of systematic, deliberative thinking (see also Chapter 2). Additionally, there's a risk of groupthink – an inclination toward premature agreement that can further narrow perspectives. The key challenge during critical deliberations and decision-making is to steer clear of these pitfalls.

To tackle this issue, Stern et al. (2020) recommend that leaders should:

- enable themselves and their team members to implement an "emergency brake". This means leaders and members should have the authority to vigorously challenge the group whenever they sense that the discussion is veering toward an incomplete, self-serving, catastrophized, or outdated understanding of the situation, or when there's a movement toward unworkable or unethical policies.

By taking this proactive stance, leaders can help ensure a more rigorous and ethical approach to decision-making during crises.

Challenge 3: Micropolitics

Stern et al. (2020) highlight that when crisis teams mobilize, political microclimates emerge that significantly influence their performance (see also Chapter 18). These climates are influenced by the quality of pre-existing relationships among the political, bureaucratic, and professional constituencies that team members represent. During a crisis, existing fault lines and competitive instincts can become even more pronounced.

According to Stern et al. (2020), to effectively manage this dynamic, leaders should:

- **Acknowledge Micropolitics:** Rather than ignoring or wishing away these political dynamics, leaders should openly address them from the outset. Naming these issues can help bring them into the light rather than allowing them to fester in the background.

- **Implement "Multiple Advocacies":** Leaders should encourage a practice of "multiple advocacies", allowing representatives of various viewpoints on critical strategic issues to present their arguments during meetings. This structured process should take place within a framework where team members remain committed to supporting decisions once they are made.
- **Act as "Magistrates":** In this context, the leader's role is to serve as a "magistrate", ensuring the integrity of the decision-making process. They should establish clear temporal or other boundaries to facilitate timely and effective decision-making.

By naming up these political micro-climates and acting as a coordinator of multiple constituencies, leaders support purposefully structured advocacy, thereby enhancing a more cooperative outcome.

Challenge 4: Tunnel Vision and Mental Fatigue

High performance often operates on the edge of failure. Sometimes stress can enhance the ability of practitioners to perform and even thrive, but it can also produce a false sense of resilience, leading teams to believe that they can simply push through (see also Chapter 5). To mitigate the risks associated with such stress, it is crucial for leaders to implement the following strategies:

- **Incorporate Reflection Moments:** Leaders should integrate regular, short intervals for teams to pause, reflect, and reconsider their approach – around 20 minutes can be effective. These moments promote mindfulness and critical thinking.
- **Establish Fallbacks and Deputies:** Designate team members to attend meetings alongside primary participants and to act as Deputies if necessary. This practice helps balance workloads, preventing the illusion of irreplaceability and the misconception that one must always be present to be valued.
- **Encourage Regular Interactions with Diverse Voices:** Team members should be encouraged to step out of isolation – whether from executive offices or confined environments – and engage with a variety of stakeholders, including frontline staff, citizens, and loved ones. These "getting out of the bunker" interactions can provide valuable perspectives and foster empathy.
- **Invite a Respected Outsider for Critique:** Bringing in an external perspective, such as a senior colleague, independent expert, or experienced facilitator, can serve as a valuable challenge to the existing beliefs, norms, and practices within the team. This "critical friend" approach, while it may seem like a luxury, is essential for healthy growth and self-awareness.

By embracing these strategies, leaders can create a more resilient and adaptable team environment, ultimately enhancing both performance and well-being.

Overcoming Mind Traps in Uncertainty

The challenges discussed above can beset leaders in VUCA-related environments because the pressures of the moment lead them to fall into mind traps. Jennifer Berger (2019) in her book *Unlocking leadership mind traps: How to thrive in complexity* suggests we each fall into several predictable mind traps that include:

- Being trapped by simple stories (we create simple characters in our stories and select data to support our beliefs).
- Being trapped by our own rightness (because it feels right doesn't mean it is).
- Being trapped by agreement (longing for agreement robs us of good ideas).

Mind Trap #1: Trapped by Simple Stories

If you have reached a conclusion about why people are behaving the way they are, for example, as a practitioner, you can crucially evaluate the degree to which you have fallen into the trap of creating a simple story. Berger's key to avoiding being trapped by simple stories is to ask yourself what other roles the actors in our stories might be playing and to come up with three additional scenarios. That is, we can disrupt the simple storyline we have created by developing the habit of carrying multiple stories about events in our lives. It is not about trying to avoid telling stories – but rather about not being trapped by one that is simplistic and one-dimensional.

Mind Trap #2: Trapped by Rightness

Being right is both seductive and dangerous. If we think we are right, then we are unlikely to be curious or open to anything that might prove us wrong. Feeling right isn't always wrong, but it is important to be open to testing by asking *what do I believe and how could I be wrong*? In conversation with others, for example, ask yourself, *Have I been listening to win*? (selectively looking for data that proves my rightness); *Have I been looking to fix*? (have I jumped to a solution to fix someone's concern?).

Mind Trap #3: Trapped by Agreement

We long for agreement because it gives us a sense of connection. We see agreement as a virtue, as an indication of our solidarity and cohesiveness. However, agreeing with what others say without consideration or withholding contradictory information is also called "social loafing". According to Berger (2019), to get along with others, we first try to compromise. When that fails, we typically polarize.

Strategies for Practitioners in Identifying Mind Traps

Consider how, in your leadership, you might have fallen into Berger's mind traps with Tool 7.1.

TOOL 7.1 THREE MIND TRAPS LEADERS CAN FALL INTO

Mind traps	Notes
Have I created a simple story? Consider a character in a scenario you have recently experienced. View the scenario as a storyline you have created. What was the story you created of this person? Create two more imaginary stories about their lives and why they acted the way they did in your scenario.	
What do I believe is the right course of action, and how could I be wrong? What have I been listening for? Might I have been listening to win or listening to fix? Where have I missed the opportunity to listen to learn someone else's perspective? Do I feel my perspective has not been listened to?	
What have we agreed on too quickly? Does this mean we have overlooked alternative perspectives that might be a better fit for purpose? How might we have listened to learn and fully understand another's perspective? What has not been said that might be getting in the way? Where were the opportunities to disagree that could lead us to expand our thinking about the issue at hand?	

Source: Berger (2019).

Overcoming these mind traps requires curiosity and a shift in the mindset being used. Karl Weick (2001b) calls this shifting from a "decision-making mindset" to one of sense-making, which is particularly important in dynamic VUCA-related environments.

Changing the Mindset to Sense-making

Karl Weick (2001b) talks about the importance of not getting "caught up" in owning one's decision (and we would argue, one's own narratives or storylines). He makes a clear distinction between sense-making and decision-making by using a quote from US Firefighter Paul Gleason, whom Weick (2001b, p. 26) talked to during a research interview. Gleason said:

> If I make a decision, it is a possession, I take pride in it, I tend to defend it and not listen to those who question it. If I make sense, then this is more dynamic, and I listen and I can change it. A decision is something you polish. Sense-making is a direction for the next period.

TABLE 7.1
Distinguishing between Decision-Making and Sense-Making

Decision-making	VS	Sense-making
Certainty	VS	Possibility
"I own my decision"	VS	"What's going on here?"
Justification	VS	Curiosity
Defending	VS	Inquiry/Updating
Or get the ego out of it!		

Source: Weick (2001b).

Our challenge then is to distinguish the difference between the mindsets of decision-making and sense-making and be conscious of which mindset is driving our thinking at that time. Table 7.1 shows some of the characteristics of a decision-making mindset compared to characteristics of a sense-making mindset.

In terms of the importance of sense-making, Karl Weick (2001a, p. 98) notes:

> The effective leader is someone who searches for the better question, accepts inexperience, stays in motion, channels decisions to those with the best knowledge of the matter at hand, crafts good stories, is obsessed with updating, encourages improvisation, and is deeply aware of personal ignorance. People who act this way help others to make sense of what they are facing. Sense-making is not about rules and options and decisions. Sense-making does not presume that there are generic right answers about things like taking risks or following rules. Instead sense-making is about trying to stay in touch with context.

As can be anticipated, when leaders operate from a place of intellectual humility, they are more likely to be able to recognize the hubris associated with "owning" one's decision as a matter of competence and instead have the self-awareness to draw instead on a mindset of sense-making. Next time you are in a situation of making a decision, do a quick reflective check which mindset you might be operating from.

MAIN POINTS

- Self-awareness is a critical skill for leaders navigating in VUCA environments.
- Intellectual humility, characterized by quiet confidence and a willingness to seek others' input, is vital for overcoming mind traps and fostering a collaborative environment.
- By acknowledging challenges such as overbearing leadership, blind spots, biases, and micropolitics, leaders can adapt their strategies to promote inclusive engagement, deliberative decision-making, and empowered team and stakeholder contributions.

- Leaders need to be mindful of not falling into the mind traps of: creating simple stories; being trapped by thinking they are right and by agreeing with others too quickly.
- One strategy that can assist here is being mindful about the differences between a sense-making and a decision-making mindset.

REFERENCES

Berger, J. G. (2019). *Unlocking leadership mind traps: How to thrive in complexity.* Stanford University Press.

Black, J., Lucas, R., Kennedy, J., Hughes, M., & Fine, H. (2024). *Command and control in the future: Concept paper 1: Grappling with complexity.* Rand Corporation. Retrieved April 28, 2025, from https://www.rand.org/pubs/research_reports/RRA2476-1.html

Heifetz, R. (1994). *Leadership without easy answers.* Harvard University Press. https://doi.org/10.4159/9780674038479

Heifetz, R., & Linsky, M. (2017). *Leadership on the line: Staying alive through the dangers of change.* Harvard Business Press.

Heifetz, R. A., Grashow, A., & Linsky, M. (2009). *The practice of adaptive leadership: Tools and tactics for changing your organization and the world.* Harvard Business Press.

Jordan, J., Wade, M., & Teracino, E. (2020). *Every leader needs to navigate these 7 tensions.* Retrieved November 24, 2024, from https://hbr.org/2020/02/every-leader-needs-to-navigate-these-7-tensions.

Marcus, L. J., McNulty, E., Dorn, B. C., & Goralnick, E. (2014). *Crisis meta-leadership lessons from the Boston marathon bombings response: The ingenuity of swarm intelligence.* National Preparedness Leadership Initiative. Harvard School of Public Health and the Harvard Kennedy School of Government, Centre for Public Leadership.

Owen, C., Hatton, T., Mitchell, J., & Parsons, D. (2025). From hero to host: Moving beyond gendered stereotypes in emergency management leadership. *The Australian Journal of Emergency Management, 39*(3), 66–72.

Owen, C., Scott, C., Adams, R., & Parsons, D. (2015). Leadership in crisis: Developing beyond command and control. *Australian Journal of Emergency Management, 30*(3), 15–19.

Starr, D. (2020). *What should crisis leadership look like?* Retrieved April 28, 2025, from www.newyorker.com/culture/annals-of-inquiry/whatshould-crisis-leadership-look-like

Stern, E., 't Hart, P., & McConnell, A. (2020). *Leading in a crisis: How to avoid common pitfalls and build a high performing team.* Australian and New Zealand School of Government. Retrieved April 28, 2025, from www.anzsog.edu.au/resource-library/resources-tlss/leadingin-a-crisis

Weick, K. E. (2001a). Leadership as the legitimization of doubt. In W. Bennis, G. M. Spreitzer, & T. G. Cummings (Eds.), *The future of leadership: Today's top leadership thinkers speak to tomorrow's leaders* (pp. 91–102). Jossey-Bass/Wiley.

Weick, K. E. (2001b). *Making sense of the organization.* Blackwell Business.

Part 2

The Social Context

In Part 2, the social context necessary to interpret and manage VUCA-related events is discussed. This invariably involves other people. Often managing high-consequence VUCA-related events involves working in or leading a team. Chapter 8 highlights some key considerations important to enhance team effectiveness. One of the critical aspects to support sharing situation awareness in times of volatility and uncertainty is a concept called psychological safety. Chapter 9 is devoted to how team leaders may encourage members to feel psychologically safe to speak up about concerns, mistakes, needing help or having an alternative perspective based on intuition or information not well known to others.

Chapter 10 takes a different approach to considering team effectiveness by exploring why sometimes teams are dysfunctional. Strategies practitioners can use to overcome some common dysfunctions are included.

The remaining chapters in this part of the book take up moving beyond the team the practitioner works within to consider other stakeholder perspectives. Chapter 11 explores how practitioners can use sense-making and narratives to increase their influence with others who may hold diverse perspectives and who need to be engaged for a cooperative effort to be achieved. Chapter 12 continues this discussion to provide some strategies for leveraging networks and alliances. To be effective, it is important for practitioners to also recognize that not all stakeholders require the same amount of attention, depending on their level of interest in the management of a VUCA situation or their power to be able to influence practitioner efforts. Chapter 13 extends consideration of the social context for leaders of organizations where the crisis represents an existential threat to the organization's survival.

DOI: 10.1201/9781003566007-9

8 Leading Teams

INTRODUCTION

Leadership plays a critical role in determining whether a team will thrive or falter (Marion & Uhl-Bien, 2001). Effective leadership fosters a positive team environment, encourages collaboration, and promotes open communication, all of which are essential for achieving collective goals. Effective leaders encourage collective goals and objectives to define a vision (Heifetz et al., 2009) and to generate and maintain confidence, cooperation, and trust within the team (Vogus & Sutcliffe, 2007). They also demonstrate and encourage constructive thinking skills that allow teams to tackle problems effectively. This includes the ability to recognize the connections between multiple issues, enabling them to address these challenges simultaneously (McNulty et al., 2021; Tourish, 2019). This capacity for integrative thinking and coordinating is essential for navigating VUCA-related conditions. Conversely, a lack of effective leadership can lead to dysfunction, miscommunication, and ultimately, a breakdown in team dynamics. Therefore, understanding the impact of leadership is important for cultivating high-performing teams able to manage in pressured environments.

DEFINING TEAMS

A team is more than just a collection of people. The formal definition of a team is where individual members have specific roles/tasks and interact adaptively to complete a common objective or shared goal (Salas et al., 2006). According to Salas et al. (2006), team members typically:

- Operate with a high degree of interdependence.
- Share authority and responsibility for self-management.
- Are accountable for the collective performance.
- Work toward a common goal and shared rewards.

For a team to operate effectively, members need to have not only sound technical skills (i.e., job role or task skills) but also the capacity to cooperate and coordinate their actions with their colleagues. In essence, there are two important types of behaviors required for a team to perform well, namely taskwork and teamwork. As the name suggests, taskwork includes the knowledge and skills required to perform a job or role. Teamwork involves the capabilities, skills, and attitudes required for a member to operate effectively within a team (Salas et al., 2006).

When team members share a strong sense of mutual commitment toward a goal, this creates synergy and can generate performance greater than the sum of the individual members. Team members and team leaders might have worked together

DOI: 10.1201/9781003566007-10

previously, or they might not be familiar with one another. Sometimes crisis management teams are formed on an ad hoc basis, where team members may or may not have previously worked together.

UNDERSTANDING THE ELEMENTS OF TEAM EFFECTIVENESS

Mike Klintworth (2014) popularized the following elements as essential for building a dream team:

1. Leadership
   The leader serves to motivate, challenge, and draw out the best in others. Leaders also encourage members to self-manage and thrive.
2. Technical Skills
   While having skilled team members is essential for accomplishing tasks, it is equally important that they contribute positively to the collective goals of the team. Creating a "dream team" involves prioritizing a supportive attitude toward teamwork alongside individual talent. One must be cautious when choosing technically gifted members; it is vital not to sacrifice attitude, behavior, or character for the sake of exceptional technical skills. Allowing any of these elements to be compromised can undermine the team's cohesion and effectiveness, ultimately leading to dysfunction and disintegration from within. Balancing talent with a collaborative mindset is key to sustaining a healthy and high-performing team dynamic.
3. Purpose
   For a team to function effectively, its members must be genuinely invested in the collective common goal or purpose. This involves understanding not only their own roles but also how their contributions align with the team's objectives. It is essential that each team member recognizes the significance of their tasks within the context of the collective mission.

   A critical point to consider is the potential for practitioners to prioritize personal interests over the team's goals. Leaders should remain vigilant for signs of self-centeredness, as this can detract from teamwork and compromise the team's overall effectiveness. Fostering a shared sense of purpose and promoting a culture of collaboration can help mitigate this risk, ensuring that all members stay focused on the mission at hand.
4. Communication
   Effective communication is the key to the success of any team. This encompasses not only the willingness to consider others' ideas but also the ability to engage in respectful and effective dialogue, especially when conflicts arise. An often-overlooked component of communication is listening. Intentional listening reflects sincerity, respect, and acceptance, fostering a more inclusive atmosphere (see Chapter 6 for developing listening skills).

   Additionally, it is important to manage team dynamics by moderating individuals who tend to over-communicate or exhibit a "know-it-all" attitude. Such behavior can be detrimental to team morale, as it can undermine the contributions of others and create an environment where team members feel discouraged from sharing their perspectives. Encouraging balanced participation ensures that all voices are heard and contributions valued. Nothing is more deflating to a team than having a member who claims to have all the answers.

5. Mutual Support

   In a successful team, each member plays a role that enhances and complements the contributions of their colleagues. This collaborative spirit is rooted in mutual respect for relationships and a genuine appreciation for the thoughts and input of all team members. Equally important is the recognition that individual actions can significantly affect others; team members must be mindful of how their decisions and behaviors impact their colleagues. By fostering a culture of mutual support and understanding, team members acknowledge that success is ultimately a collective effort. This mindset encourages cooperation, collaboration, and a shared commitment to achieving the team's goals, creating an environment where everyone feels valued and motivated to contribute their best.

6. Character

   For a team to work well members need to have integrity, loyalty, and to act ethically.

7. Attitude

   According to Klintworth (2014), attitude plays a crucial role in either energizing a team or undermining its effectiveness. Negative team members can create discord, often influencing their colleagues to adopt a similarly destructive outlook, which is detrimental to the team's health and productivity.

In the book *The Culture Code*, by Daniel Coyle (2018), he speaks about the kinds of practices team leaders can engage in to support cooperative member behaviors. These are discussed further in Chapter 9. Also useful for team leaders in addressing negative attitudes is an understanding of team dysfunction (see Chapter 10).

STRUCTURE AND CULTURE OF TEAMWORK

Richard Hackman's (1983) initial research identified five key conditions that enhance teamwork and increase the likelihood of effectiveness:

1. *Being a Real Team:* Effective teams are characterized by a clear boundary that defines who is part of the team. Members are interdependent, and there is a moderate level of stability in membership, which fosters consistency and collaboration. We argue this can also be the case in the emergent teams described by Edmondson and Lei (2014) because being part of a team includes members fulfilling particular roles, even if for a short time.
2. *Having a Compelling Direction:* Successful teams work toward clearly defined, challenging goals that carry sufficient importance to motivate

members. A compelling direction inspires collective effort and aligns individual contributions with the team's objectives.

3. *Having an Enabling Structure for Teamwork:* The team's structure – how it organizes and conducts tasks – should facilitate teamwork rather than hinder it. For example, if decision-making power is concentrated in one individual while a large number of people contribute work, this hierarchical structure can obstruct the team's overall effectiveness.
4. *Having a Supportive Context:* A supportive organizational environment is crucial for a team's efficiency. This includes access to adequate resources, appropriate rewards, relevant information, and the cooperation and support of the organization to enable the team to perform effectively.
5. *Having Expert Coaching and Guidance:* Access to coaching or mentorship can significantly benefit teams. Having an expert available to provide guidance and help navigate challenges can enhance the team's performance and development.

These conditions underscore the importance of a well-defined, supportive, and collaborative environment for effective teamwork. Hackman (2011) later added a 6th dimension (Clear Norms), which is also included in Tool 8.1. Tool 8.1 is a team assessment tool based on Hackman's (1983, 2002, 2011) model of team effectiveness.

ENABLING CONDITIONS FOR TEAMS

Assessing the Enabling Conditions of a Team

You can use the Tool to self-assess how the team you lead performs on the six key enabling conditions Hackman identified. Review the items in the Tool and give your team a score from excellent to poor (A, B, C, D or E) for each of them. Reviewing your own self-assessment may give you some ideas about specific areas to address to lift the team's effectiveness (further discussed in Chapter 9).

Launching a Team for Success (the 60/30/10 Rule)

According to Hackman (2011), much of what a member of the team brings to their role has already been previously established (60%) before they walk into the room. What a leader can influence in terms of their capability is setting the tone (30%) of how they operate as part of the team, and what can be improved in coaching and mentoring (10%).

There are some key things leaders can do to launch a team for success. The following were identified by Hackman's (2011) involvement with Project Looking Glass, a simulation following the 9/11 terrorist attack on the World Trade Centre. This project identified that the way people in a new team were introduced had a significant impact on the team's effectiveness. It was found that introduction by job title and job status inhibited openness and curiosity about what skills

TOOL 8.1 TEAM ENABLING CONDITIONS ASSESSMENT

A	B	C	D	E
Excellent		So-so		Poor

Key enabling conditions	Score
Real team	
• It is clear who is and who is not a member?	○
• Members are interdependent for accomplishing the work.	○
• The team has reasonable stability of membership.	○
Compelling purpose	
• Team purposes are clearly stated.	○
• The team's purposes are a "stretch" to achieve.	○
• The team's work has consequences for others.	○
• The team's purposes specify the ends sought but not procedural details.	○
Right people	
• Members have ample task and teamwork skills.	○
• There are no "derailers" on the team.	○
• The team has the right number and mix of members.	○

Key enabling conditions	Score
Clear norms	
• Team norms of conduct are clear and appropriate.	○
• Norms support the full use of members' contributions.	○
• Norms support active planning of team performance strategies.	○
Supportive context	
• Information needed for the work is readily available.	○
• Educational and technical assistance are available when needed.	○
• Good team performance is recognized and reinforced.	○
• The team can get the material resources it needs for the work.	○
Available coaching	
• The team has ready access to coaching assistance.	○
• Coaches are competent in dealing with team dynamics.	○

Source: Hackman (2011).

members had that could contribute to problem-solving. Teams that introduced themselves by the skills and knowledge they brought to the table (i.e., identified what they could contribute to the team for success) were ultimately more effective at working together.

These teams utilized everyone's abilities to achieve goals. Solving complex problems in rapidly developing VUCA contexts requires effective exchange of information between participants, the generation of solution options, and the critique of possible solutions from multiple perspectives. Items such as uniforms, job status, and the way introductions are done can inhibit effective information flows, solution generation, and identification of problems and opportunities.

STRATEGIES FOR PRACTITIONERS

To launch a team for success, Hackman (2011) proposed that the leaders could adopt the following practices at the outset of:

- Set team climate
  - Encourage challenge
  - Set the power distance (that is lower the hierarchical power-authority gradient (see Chapter 9)
  - Set desired behaviors
- Clarify goals
- Create a team knowledge pool
- Connect boundaries
- Encourage members to test assumptions

Other Challenges to Team Effectiveness

While it is often the case that there can be challenges that need to be overcome for team members to speak up, sometimes there can be members who love the sound of their own voice. Team leaders need to monitor how much and what contributions all team members are making and beware of the member who over-communicates; that is, those who appear to be the "know it all".

Most often, however, the opposite is the case – members notice things but don't feel compelled to share them. This is because there is an assumption that if one team member has observed or thought about something, then, surely other members (potentially with more experience) have as well. This is known as "false consensus". Leaders of teams need to create a climate where members can all share their resources and knowledge.

When team members contribute to a complex problem, they help in increasing adaptability and flexibility because more options are available. These include alerting others that something has changed in a situation and coming up with multiple plans to resolve a problem that can then be considered. Team leaders need to encourage team members to contribute to alternative solutions. Some ways in which practitioners can encourage engagement include legitimizing doubt and articulating uncertainty to encourage input from others.

Legitimizing Doubt and Articulating Uncertainty

Articulating uncertainty in VUCA-related environments does two things. First, it alerts others to what they might need to be vigilant for, and second, it gives other members a voice to speak up and articulate their uncertainty. Some fear that doing so will lead to paralysis, but under most circumstances, the opposite is the case. One approach that supports encouraging people to speak up is one developed by Weick and Sutcliffe (2007) called the STICC (Situation, Task, Intent, Concern, Calibrate) Protocol. This is outlined in Table 8.1

As was discussed in Chapter 2, when under pressure, the thinking process typically narrows and our brains seek obvious black-and-white, yes-or-no types of answers to return to the security of certainty. This means it is important for leaders to use strategies that open up curiosity and explore differences in perspectives and even dissent (Marquardt & Tiede, 2023).

One way of doing this is for leaders to express uncertainty themselves and be curious about what the alternatives might be. According to Duke (2018), leaders who acknowledge their uncertainty foster better truth-seeking within teams and encourage the sharing of valuable information and dissenting opinions. The fear of being "wrong" or pointing out someone else's errors often conflicts with the unspoken social contract of confirmation, causing individuals to withhold critical insights.

By openly expressing uncertainty, a leader creates a space where team members understand that discussions are not about assigning right or wrong, but rather about exploring ideas collaboratively. Duke (2018) suggests that acknowledging uncertainty enhances credibility as a communicator. While it is easy to fall into the trap of believing that unwavering confidence is necessary to gain respect, acknowledging uncertainty can actually strengthen trust and dialogue.

Clearly, there is a balance required between expressing uncertainty in life-threatening high-consequence environments. That is, it is important not to look too nervous or unsure. However, a credible leader will guide team members to seek truth rather than jumping to the first thing that they think of, which is an error trap because the brain wants to reach certainty rather than dwell in complexity or chaos.

TABLE 8.1
STICC Protocol

Situation →	**Task →**	**Intent →**	**Concern →**	**Calibrate**
Here's what I think we face	*Here's what I think we should do*	*Here's why I think this is what we should do*	*Here's what we should keep our eye on because if that changes, we're in a whole new situation*	*Now talk to me. Tell me if you don't understand, cannot do it, or see something I do not*

Source: Weick and Sutcliffe (2007, p. 156).

One way of doing this is to use probability when communicating – and seeking – levels of confidence. For example, if someone stated, *I believe this to be true and I'm 80% confident about it*, then it invites self and others to explore the quantity and quality of the information with thoughtfulness and self-awareness. Thoughtful and self-aware people are more credible. Expressing levels of confidence also invites team members to become collaborators. As indicated in Tool 8.2, encouraging team members to use probability to express their confidence can assist in clarity and also identify degrees of doubt (e.g., *I think this plan will work, and I'm 80% confident on it*) illustrates the level of confidence and invites members to consider what might be the 20% of doubt, including having a discussion about how to develop contingency plans to shore up the uncertainty.

TOOL 8.2 EXPLORE PROBABILITY – WHAT IS THE LIKELIHOOD?

1. Think of a future decision you will be making and pose it as a question using probability-based questioning. For example, "What's the probability (or likelihood) that ….?"
2. Circle the probability level below that you think it will be the likely outcome.

0%	5%	30%	50%	80%	95%	99%	100%

David Marquet (2015) in his book *Turn The Ship Around!* would use something similar in his team development meetings. He would ask a question (e.g., *What is the likelihood that this project plan will be a success?*) and then encourage team members to throw in a probability card. He would then turn these over and then look to the outliers and say *What would someone who is thinking 30% or 80% be seeing?* By doing so, he enables members to safely explore differences in perspective.

MAIN POINTS

- Leadership plays a critical role in determining whether a team will thrive or falter, fostering a positive environment, collaboration, and open communication.
- A team is a collection of individuals with specific roles/tasks who interact adaptively to complete a common objective or shared goal, operating with interdependence, shared authority, and collective accountability.

- For a team to perform well, members need both **Taskwork** (technical skills for their job/role) and **Teamwork** (capabilities, skills, and attitudes for effective team collaboration).
- Teamwork and taskwork skills that emphasize clarity in goals and role responsibilities set up opportunities to enhance team dynamics.
- Effective leaders define a collective vision, generate trust, demonstrate constructive thinking skills, and act as emotional shields in times of crisis to manage negative emotions and stress.
- Essential elements of team effectiveness (as per Mike Klintworth, 2014) include: 1. Leadership 2. Technical Skills (balanced with a collaborative mindset) 3. Purpose (shared common goal or mission) 4. Communication (effective, respectful, and inclusive).
- A variety of strategies can be used to articulate uncertainty and to legitimize doubt in the spirit of sense-making. This includes using probability-based questioning.

REFERENCES

Duke, A. (2018). *Thinking in bets: Making smarter decisions when you don't have all the facts*. Portfolio Press.

Edmondson, A. C., & Lei, Z. (2014). Psychological safety: The history, renaissance, and future of an interpersonal construct. *Annual Review of Organizational Psychology and Organizational Behavior, 1*(1), 23–43.

Hackman, J. R. (1983). *A normative model of work team effectiveness*. Report for the Office of Naval Research, Arlington, VA.

Hackman, J. R. (2002). *Leading teams: Setting the stage for great performances*. Harvard Business Press.

Hackman, J. R. (2011). *Collaborative intelligence: Using teams to solve hard problems*. Berrett-Koehler Publishers.

Heifetz, R. A., Grashow, A., & Linsky, M. (2009). *The practice of adaptive leadership: Tools and tactics for changing your organization and the world*. Harvard Business Press.

Klintworth, M. (19 September 2014). *The 7 essentials for building a dream team*. Retrieved January 9, 2020, from https://generalleadership.com/7-essentials-building-dream-team/

Marion, R., & Uhl-Bien, M. (2001). Leadership in complex organizations. *The Leadership Quarterly, 12*(4), 389–418.

Marquardt, M., & Tiede, B. (2023). *Leading with questions: How leaders discover powerful answers by knowing how and what to ask*. John Wiley & Sons.

Marquet, L. D. (2015). *Turn the ship around!: A true story of building leaders by breaking the rules*. Penguin Books Limited.

McNulty, E., Marcus, L., Grimes, J., Henderson, J., & Serino, R. (2021, November 29). The meta-leadership model for crisis leadership. *Oxford Research Encyclopedia of Politics*. Retrieved November 8, 2025, from https://oxfordre.com/politics/view/10.1093/acrefore/9780190228637.001.0001/acrefore-9780190228637-e-2032

Salas, E., Rosen, M. A., Burke, C. S., Goodwin, G. F., & Fiore, S. M. (2006). The making of a dream team: When expert teams do best. In K. A. Ericsson, N. Charness, P. J. Feltovich, & R. R. Hoffman (Eds.), *The Cambridge handbook of expertise and expert performance* (pp. 439–453). Cambridge University Press. https://doi.org/10.1017/CBO9780511816796.025

Tourish, D. (2019). Is complexity leadership theory complex enough? A critical appraisal, some modifications and suggestions for further research. *Organization Studies*, *40*(2), 219–238.

Vogus, T. J., & Sutcliffe, K. M. (2007). The impact of safety organizing, trusted leadership, and care pathways on reported medication errors in hospital nursing units. *Medical Care*, 997–1002.

Weick, K. E., & Sutcliffe, K. M. (2007). *Managing the unexpected* (Vol. 9).

9 Team Effectiveness and Psychological Safety

INTRODUCTION

The previous chapter identified the structural and cultural attributes required for effective teamwork and defined a team as an interdependent unit where members have specific roles and interact adaptively, emphasizing the importance of both taskwork and teamwork skills. It highlighted the necessity of clarifying goals, creating a knowledge pool, and testing assumptions to enhance team performance. In a literature review undertaken by Collins et al. (2023), working in crisis and adversity was highlighted as being stressful for both leaders and their teams. Team members frequently look to their team leaders to alleviate stress and manage negative emotions. This observation has implications for leaders to act as emotional shields for their members, protecting them from high negative emotion and arousal. This also means that leaders in these situations need the awareness and skills to manage high arousal negative emotions (see Chapter 5).

Sometimes we don't have the luxury of working in a stable team and instead might be working with others in what Edmondson and Lei (2014) have called emergent teams or "teamwork on the fly". Unlike stable teams (where people have stable, bounded formal roles in the way described in Chapter 8), emergent teams bring together people with different expertise to form temporary teams to solve a particular problem. In this respect, people who do not know one another come together to work on a shared goal. We believe that this kind of teaming is becoming increasingly prevalent and more important as we face novel problems. Under these conditions, understanding the leadership qualities to support psychological safety by empowering people to speak up is paramount.

This chapter discusses the significance of managing team discussions, encouraging open communication, and fostering a climate where team members feel comfortable with challenging perspectives. This is because effective teamwork is achieved through human processes and relationships between members (Barton & Sutcliffe, 2009).

PSYCHOLOGICAL SAFETY AND VOICE

One of the key differences between an average team and one that performs effectively is a concept called psychological safety (Edmondson, 1999, 2004). Research has identified (Edmondson & Bransby, 2023; Edmondson, 1999, 2005) the critical role of team psychological safety in fostering an environment where team members feel comfortable speaking up and offering constructive feedback, including dissent. Psychological safety develops through shared experiences and involves building trust and mutual respect within the team (Edmondson, 1999).

DOI: 10.1201/9781003566007-11

When a team has developed psychological safety, members are confident they won't be undermined, penalized, or embarrassed by their colleagues for speaking up, asking questions, proposing new ideas, reporting errors, or questioning a decision (Edmondson & Lei, 2014; Edmondson et al., 2001). Additionally, psychological safety supports team learning and provides ongoing opportunities to enhance team performance (Edmondson et al., 2001).

Members provide and seek information to make meaning of the cues that they think are important (which include uncertainty and discomfort), and the team adjusts, tweaks, and adapts to this information flow, which if not forthcoming could result in larger problems and potential failures. Later in this chapter, we will explore some of the strategies team leaders can use to facilitate speaking up and voice in support of enhanced situation awareness.

Inter-Agency and Emergent Teams

The diversity of inter-agency and emergent teams assembled for high-stakes VUCA situations presents a particular challenge for leaders (Edmondson & Lei, 2014). Developing psychological safety in such teams is likely to be more difficult. Edmondson (2004) suggests that organizations can foster psychological safety by enabling personnel to practice together in simulated exercises (i.e., "practice fields" – Senge, 1990) and through effective leadership behavior. Leaders can support team psychological safety by:

1. making themselves available and approachable
2. explicitly asking for input
3. modeling openness and fallibility.

Strategies to cultivate psychological safety include ensuring clarity about team member job responsibilities and support from colleagues. This is because the capacity to feel confident about speaking up involves clarity about what is expected of the roles each team member plays. This is especially important when there is a high degree of interdependence between team member roles (Hirsch, 2017).

Building Psychologically Safe Environments

In drawing off the work of Edmondson and Bransby (2023), Ronnie Kinsey (2024) outlines the key practices team leaders can use to enhance psychological safety.

1. **Normalize trust and vulnerability.** Leaders can do this by expressing uncertainty and asking for input.
2. **Be focused on solutions.** Errors will happen and when they do it is important to avoid the blame game and instead work toward finding a solution.
3. **Engage in active listening.** This can be as simple as recapping what you heard someone say before responding. The importance of active listening has also been discussed in Chapter 6.

4. **Normalize failure.** When failures do happen, ensure they are treated as an opportunity for learning and growth.
5. **Seek different perspectives.** Actively solicit opinions from the less vocal.
6. **Be open and approachable.** This can be as simple as when a team member visits your office, closing your laptop so they understand they have your full attention.
7. **Consistent check-ins.** Ensure you follow up with members to learn how they're progressing.
8. **Address issues promptly.** When issues arise engage them quickly.

Assess Your Team's Psychological Safety

In the Psychological Safety Playbook developed by Geraghty (2020), there are ten questions that can be used to gain feedback on how team members feel about the team communication climate. Use Tool 9.1 to assess your own sense of psychological safety within a team. You could also ask the members of your team these questions in an anonymous survey. This would give the practitioner insight as to where they might need to focus attention to enable a more open communication climate. Team leaders can then have conversations with members to highlight and visibilize any areas of concern about team functioning.

Overcoming Misconceptions About Psychological Safety

It is important to acknowledge that psychological safety is not just about people being comfortable. It is instead about having open conversation with what Edmondson and Kerrissey (2025) call "candor". They note several misconceptions about what psychological safety is and point out that it is not about being "nice". Instead, it is about feeling safe enough to give clear feedback if a member's work has fallen short and being as constructive as possible (see also Chapter 10). In teams where there is no disagreement there is a risk of mediocracy. Edmondson and Kerrissey (2025) propose that to build psychological safety leaders need to:

- Emphasize work goals and purposes and why they matter. In asking questions of team members such as "*what do our stakeholders want or need from us*" and "*what will it take to deliver that*?" "*Who has a different view*?" "*If this decision backfired, what would turn out to be the reason?*" (See also conducting a pre-mortem in Chapter 15 *Foresight for Futures Thinking*). These conversations make it easier for members to report bad news or express a dissenting view.
- Improve the quality of team conversations. Leaders can role model candid sharing that is processed thoughtfully. Skills in asking good questions and listening intently have already been discussed in Chapters 6 and 11.

Tool 9.2 provides a guide to how leaders can assess the quality of their team conversations.

TOOL 9.1 ASSESS YOUR PSYCHOLOGICAL SAFETY

	Self-review quiz	Strongly disagree	Disagree	Neutral	Agree	Strongly agree
1	On this team, I understand what is expected of me	O	O	O	O	O
2	We value outcomes more than outputs or inputs, and nobody needs to "look good"	O	O	O	O	O
3	If I make a mistake on this team, it is never held against me	O	O	O	O	O
4	When something goes wrong, we work as a team to find the systemic cause	O	O	O	O	O
5	All members of this team feel able to bring up problems and tough issues	O	O	O	O	O
6	Members of this team never reject others for being different and nobody is left out	O	O	O	O	O
7	It is safe for me to take a risk on this team	O	O	O	O	O
8	It is easy for me to ask other members of this team for help	O	O	O	O	O
9	Nobody on this team would deliberately act in a way that undermines my efforts	O	O	O	O	O
10	Working with members of this team, my unique skills and talents are valued and utilized	O	O	O	O	O

Source: https://www.psychsafety.co.uk/is-your-team-psychologically-safe-the-ten-statement-quiz/

Speaking Truth to Power

In hierarchical organizations, it is often the case that there is a need to speak up to someone in higher authority and who holds a more powerful position. The term "authority gradient" was first defined in aviation when it was noted that pilots and co-pilots may not communicate effectively in stressful situations if there is a significant difference in their experience, perceived expertise, or authority.

For example, on January 13, 1982, an Air Florida aircraft crashed into the Potomac River due to excessive snow and ice on the aircraft and a frozen indicator which gave the crew a false engine power reading (Fischer & Orasanu, 1997). The

First officer: Look how the ice is just hanging, ah, back there, see that?
(…)
Captain: yeah
After a long wait following de-icing, the first officer continued:
First officer: Boy, this is a, this is a losing battle here on trying to de-ice those things, it (gives) you a false feeling of security, that's all that it does.
Shortly after being given clearance to take off, the first officer again expressed his concern:
First officer: Let's check those tops again since we've been sitting here awhile.
Captain: I think we get to go here in a minute.
Finally, while they were on their takeoff roll, the first officer noticed that something was wrong with the engine readings.
First officer: That don't seem right, does it? [three second pause] Ah, that's not right.
Captain: Yes, it is, there's 80.
First officer: Naw, I don't think that's right. [seven second pause] Ah, maybe it is.
Captain: Hundred and twenty.
First officer: I don't know.

Source: Fischer and Orasanu (1997).

following box is a transcript taken from the cockpit voice recorder of how the first officer attempted to bring his concerns to the attention of the inattentive pilot.

The transcript shows how, in the face of risk, the lower ranked officer was unable to express their concerns directly and instead used veiled comments. This indicates a need, especially in hierarchical teams, to provide support to lower ranked members to speak up about their concerns.

Using the Graded Assertiveness Protocol

In some domains, a communication skill to help people challenge and speak up to overcome the "power-authority gradient" has been called the graded assertiveness protocol.

According to the Emergency Department Clinical Facilitator (Anonymous, 2014), graded assertiveness is a learned skill and having a template to follow encourages safety to practice. It is a process of communicating, advocating, and directing with hard-core clarity that is useful in stressful or crisis scenarios.

During high-pressure events, there are a number of factors that can inhibit communication which include differences in seniority. This can lead to people not speaking up for fear of losing face or not being seen as decisive. For example, in acute hospital care environments, one acronym that has been developed to support members of teams who are of lower rank (nurses) is that of PACE. The four levels included in PACE are steps to escalate a concern if the problem is not resolved.

TABLE 9.1
Strategies to Speak up to Authority – Graded Assertiveness in Healthcare

Level	Message
Probe/Non-verbal behavior	I notice that the dose is… / Are you sure the dose is…?
Alert/Offer alternatives	Would you like me to help with… Shall we check the MIMS?
Challenge/Ask for clarification	Is there a reason you're going to double the dose? I think the dose is actually… / I'm concerned about…
Emergency action	Doctor, you must listen! That dose is incorrect for this patient. I'm not going to give it. I'm calling the registrar now.

Source: www.Theemergencydeptclinicalfacilitator.Wordpress.com/2014/06/03/pace-yourselves

In this way, the power to resolve the problem transfers from the *other*, to *shared* and finally to *self*. Table 9.1 highlights examples of each of the levels of escalation.

A key attribute of a strategy like PACE is that it also puts the onus of responsibility on both parties. In the first instance, the onus is on the lower-ranked person not to simply give up after expressing one indirect concern if that concern is dismissed. In the second instance, the onus is also on the higher-ranked person to be attentive to the expressions of concern being repeatedly shared.

In the professional development work leading to this book, the first author (Owen) has used a similar approach with incident or emergency management teams to explore awkward conversations; by using the "thinking one thing but saying another" cartoon illustrated in Figure 9.1.

Individuals were asked to think of an experience where they had an "awkward moment" and how, in using the steps listed in the graded assertiveness protocol, the expression of concern could have been appropriately escalated. Participants were also invited to color in the tabards which indicate the functional team roles held if relevant to their case example. The example in Figure 9.1 illustrates a peer-to-peer conversation between an Incident Management Team Planning Officer and an Operations Officer and represents an event that actually happened. In this example the Planning Officer thinks "*The ICC is in the path of the fire. We are going to die*" but instead they say to their colleague "*As a result of the fire behaviour analysts' and weather predictions, what are your thoughts on moving the ICC?* to which the Operations Officer brushes off the expression of concern by replying "*We'll have the fire under control by then, and it's too much of a hassle to shift now!*" The use of PACE by the Planning Officer could escalate their concern through: Level 1 Showing their colleague a predictive map of anticipated fire trajectory; and/or opening the window to show the amount of smoke and its direction. The rest of the figure illustrates the way in which the concern could have been escalated ultimately declaring the position of the center is unsafe and calling on a higher authority.

In the first author's experience, many of the examples provided through undertaking the "awkward moment" exercise in professional development resulted in participants narrating real events. In this case, the incident control center did have to be

Levels	Messages
Probe/ Non-verbal behaviour	Shows Ops predictive map. Opens the window
Alert /offer alternatives	What are our contingences then?
Challenge/ask for clarification	Why do you think it is too hard? What are the alternatives?
Emergency Action/Intervention	We need to shift the ICC now- it is unsafe! I'm calling the Commander

FIGURE 9.1 The PACE protocol used in professional development with emergency incident management team members

Source: **First Author (Owen).**

relocated, and at the worst possible moment, a theme often mentioned in Coronial inquiry reports and other Commissions of Inquiry into management of such events.

A content analysis conducted of a sample of 25 "awkward moments" was also insightful. It revealed that most of the examples pertained to practitioners "managing up", that is in attempting to express concern to an authority and where the concern represented a refusal to undertake action wanted by the boss. What was also troubling was that in more than half the cases, the challenge was still not explicitly articulated to address the emergency action required. For example, "*I'm not going to do it – do it yourself*" versus "*That is unsafe. I am not taking my crews in there*".

In a nutshell, psychological safety must be cultivated. Research with US wildland firefighters, for example, revealed that even experienced and senior personnel sometimes face social pressures that inhibit them from speaking up in critical situations (Lewis et al., 2011). This speaks of an organizational culture that often pervades traditionally hierarchical organizations where there is a stronger deference to authority then there is to situational expertise (Owen, 2014).

Assessing Conversation Quality for Psychological Safety

Barton and Sutcliffe (2009) interviewed 28 experienced firefighters and from those interviews extracted 62 cases of incidents that had either gone well or had resulted in bad outcomes. A key difference between those that ended badly and those that did not was the extent to which individuals voiced their concerns about early warning signs.

While a leader may have considerable expertise based on many years of previous experience, in VUCA situations this does not make them a situational expert who sees the same cues or understands them as understood by other members of the team. Where conditions are rapidly changing and ambiguous, constant updates and adjustments are needed. Under these conditions, the team needs to support and maintain collective situational expertise.

Leaders can assess the quality of the conversations they cultivate in their teams. Edmondson and Kerrissey (2025) have prepared a guide for leaders to assess the quality of conversations occurring in their teams that indicate whether psychological safety is present.

TOOL 9.2 ASSESSING TEAM TALK FOR PSYCHOLOGICAL SAFETY

1	2	3	4	5
Not at all		Sometimes		Frequently

Positive	Score	Negative	Score
Attribute: People are contributing and listening			
People are candid in their sharing of information and opinions?		People seem to be holding back and saying only safe things.	
Everyone appears to be engaged in the substance of the discussion.		One person or a few dominate the conversation.	
Attribute: Advocacy and inquiry are both present			
There is a good balance between advocacy (to promote certain ideas) and inquiry (to understand them).		People get stuck trying to prove their original points, regardless of new information.	
People are asking good questions that prompt others to respond to their ideas and concerns.		People are selective in their use of information to push for their position	
Attribute: Mutual learning and progress are happening			
I often feel I'm gaining a better understanding of the topic as the conversation goes on		The conversation goes around in circles.	
The conversation is disciplined, systematic, and data-driven		People just dig in and repeat their opinions	
Total for positive and negative attributes- which is higher?	Positive score		Negative score

Edmondson and Kerrissey (2025) note that psychological safety is reinforced by structures and rituals that prompt teams to talk about their progress and are honest in their appraisals. When team members engage in robust conversations, they are more likely to be reporting small or weak signals that something might be wrong (including with their own capability). This shared situation awareness enables team members to adjust to VUCA conditions by reorganizing resources and actions as necessary. Some of the reasons why robust conversations can be inhibited are discussed in the next chapter on understanding team dysfunction.

MAIN POINTS

- Psychological safety is identified as a key differentiator between average and high-performing teams, allowing team members to feel comfortable speaking up and offering feedback.
- It is essential for team leaders to cultivate psychological safety through shared experiences and building trust and respect among team members.
- Understanding the differences between stable teams and temporary/emergent teams will be increasingly important for solving novel problems.
- Effective communication within teams can be hindered by dominant voices and the false consensus effect, where practitioners assume others have noticed or thought the same things they have.
- Leaders need to create an environment that encourages open communication, allowing team members to comfortably challenge perspectives and share insights.
- The chapter introduced the PACE protocol, a graded assertiveness tool designed to assist team members, particularly in high-pressure situations, to escalate concerns effectively.
- Team leaders can self-assess the quality of their team's talk for psychological safety.

REFERENCES

Anonymous (2014, June 3). *PACE yourselves!* The Emergency Department Clinical Facilitator. Retrieved May 14, 2025, from www.theemergencydeptclinicalfacilitator.wordpress.com/2014/06/03/pace-yourselves

Barton, M. A., & Sutcliffe, K. M. (2009). Overcoming dysfunctional momentum: Organizational safety as a social achievement. *Human Relations*, *62*(9), 1327–1356.

Collins, M. D., Dasborough, M. T., Gregg, H. R., Xu, C., Deen, C. M., He, Y., & Restubog, S. L. D. (2023). Traversing the storm: An interdisciplinary review of crisis leadership. *The Leadership Quarterly*, *34*(1), 101661.

Edmondson, A. (1999). Psychological safety and learning behavior in work teams. *Administrative Science Quarterly*, *44*, 350–383.

Edmondson, A. (2004). Psychological safety, trust, and learning in organizations: A group-level lens. In R. M. Kramer, & K. S. Cook (Eds.), *Trust and distrust in organizations: Dilemmas and approaches* (pp. 239–272). Russell Sage Foundation.

Edmondson, A. (2005). Psychological safety and learning behavior in work teams. In C. L. Cooper, & W. H. Starbuck (Eds.), *Work and workers* (vol. 3(9), pp. 291–329). Sage Publications.

Edmondson, A., Bohmer, R., & Pisano, G. (2001). Speeding up team learning. *Harvard Business Review, 79*, 125–132.

Edmondson, A. C., & Bransby, D. P. (2023). Psychological safety comes of age: Observed themes in an established literature. *Annual Review of Organizational Psychology and Organizational Behavior, 10*(1), 55–78.

Edmondson, A. C., & Kerrissey, M. J. (2025). What people get wrong about psychological safety. *Harvard Business Review, 103*(3), 52–59.

Edmondson, A. C., & Lei, Z. (2014). Psychological safety: The history, renaissance, and future of an interpersonal construct. *Annual Review of Organizational Psychology and Organizational Behavior, 1*(1), 23–43.

Fischer, U., & Orasanu, J. (1997). How to challenge the captain's actions. In *International symposium on aviation psychology* (9th ed., pp. 624–628).

Geraghty, T. (2020). *Psychological safety playbook*. Retrieved May 14, 2025, from https://www.psychsafety.co.uk/is-your-team-psychologically-safe-the-ten-statement-quiz/

Hirsch, W. (2017). *Five questions about psychological safety, answered. Evidence summary.* Retrieved May 11, 2025, from www.scienceforwork.com/blog/psychological-safety

Kinsey, R. (2024). *Nine tips for leaders to create safety in teams*. Retrieved May 15, 2025, from www.business2community.com/strategy/measure-psychological-safety-team-01730787

Lewis, A., Hall, T. E., & Black, A. (2011). Career stages in wildland firefighting: Implications for voice in risky situations. *International Journal of Wildland Fire, 20*, 115–124.

Owen, C. (2014) Diagnosis in emergency management. In M. Wiggins (Ed.), *Diagnostic expertise in organisational environments*. Ashgate.

Senge, P. M. (1990). *The fifth discipline: The art and practice of the learning organization.* Doubleday.

10 Understanding Team Dysfunction

INTRODUCTION

As was discussed in the previous chapters, for a team to operate effectively, members need to have not only sound technical skills (i.e., job role or task), but also the capacity to cooperate and coordinate their actions with their colleagues. As Chapter 9 highlighted, the latter is supported by having psychological safety to speak up about concerns and to offer differing observations and perspectives. Team leaders need to build an environment where team members can communicate effectively and respectfully when conflict arises.

However, sometimes teams are not effective because they suffer from a series of common dysfunctions. It is the job of a leader to diagnose what might be holding a team back from performing at its best.

UNDERSTANDING TEAM DYSFUNCTION

In 2006, author Patrick Lencioni published his book *The Five Dysfunctions of a Team*, which outlines a model for work team effectiveness based on identifying and addressing dysfunctions. The model is one of a pyramid where one dysfunction, if not addressed, can lead to another. The base of the pyramid is about trust.

Absence of Trust: Team members need to feel that others understand and act in ways that demonstrate their actions and intentions are constructive and that all are committed to the same goals. If this is not in place, members are unlikely to feel comfortable in asking for help, admitting mistakes, or revealing uncertainty or vulnerability to others. Without trust, there's no comfort level needed to ask for help when needed, which can lead to delays and mistakes that could have been avoided with early intervention.

When team members trust one another, they can seek needed support and express uncertainty. This is the foundation of all team effectiveness and why psychological safety, discussed in Chapter 9, is so important. If there is no trust, then all other dysfunctions discussed below can be enabled.

Fear of Conflict: This is the second level that sits above a lack of trust. If everyone is trying too hard to maintain an artificial sense of peace, there won't be any dynamic discussion about differences of opinion that can lead to productive ideas. In teams where there is a fear of conflict, it is likely that members make guarded or vague comments, avoiding any debate or discussion that could lead to a critical evaluation of proposed ideas and in turn to better solutions.

DOI: 10.1201/9781003566007-12

The capacity to constructively disagree with one another is essential to success because it supports finding the best solutions to problems. If there is no trust that members come from a well-intentioned place, then key issues are not able to be openly aired.

Lack of Commitment: If team members are just simply going along with decisions because they risk an argument, they are not likely to be committed to their work or the team. Under these circumstances, decisions are paid lip service rather than members fully investing in them.

In most cases people with alternative perspectives or differences of opinion are not necessarily hell-bent on getting their idea supported, but they do value being heard and having their ideas considered. If this is possible and they feel they have been listened to, and their ideas have been given due consideration but at the end of the discussion if an alternative is chosen, then they are more likely to commit to it because they can see the team has considered and clarified what is important.

Avoidance of Accountability: When there is no clear plan of action and concerns have not been addressed, then it is likely that this fear of conflict and lack of commitment can lead to a reluctance to hold others accountable for their work. Under these circumstances, when team members, for example, fail to complete tasks on time or to the standard required no one calls them out on it. This in part is because there have been assumptions that have not been aired so they may be clarified, and there has not been a passionate debate about what is needed to get the job done.

Teams that have clear standards of performance and clarity about roles, together with psychologically safe conditions where difference can be discussed, are more likely to be able to raise performance issues of concern with one another. It is not just the team leader's role but one that is collectively shared.

Inattention to Results: If personal goals become more important than the group's success, no one will focus on collective results or plan how to improve them. Under these circumstances, team members are more likely to prioritize their individual achievements and recognition over the team's collective goals, resulting in fragmented efforts and suboptimal outcomes.

Team members who trust one another engage in passionate dialogue about what matters and can question and challenge each other and where there is clarity around responsibilities and standards of performance are more likely to set aside personal interests and support the collective team goal.

OVERCOMING TEAM DYSFUNCTION

Overcoming Lack of Trust

Teams that can trust one another perform better when they allow for questioning, raising concerns, and acknowledging mistakes. Building trust and mutual respect are enablers for psychological safety within the team (Edmondson, 1999, 2004; Edmondson et al., 2001).

Building Swift Trust

In VUCA situations, there may be little time to develop trust. The challenge is that trusting others involves openness and vulnerability, which typically requires ceding control. On his blog, *Influence*, Warren (2017) discusses three principles that support building swift trust in teams:

1. Establishing Common Ground and Shared Purpose
   Create an environment of "we, us, our". A crisis focuses attention on the task or challenge, with personal concerns set aside. When a group has a clear and shared purpose, it helps everyone focus on what matters and understand their roles and the roles of others. Seeing each other as sharing common ground – the situation isn't yours or mine, but ours – is important. This can be challenging when there is uncertainty. It is important to encourage a sense of shared purpose and mutual respect for the collective contributions of members.
2. Encouraging the Common Components of Trust
   These include openness, acceptance, congruence (alignment of what we say and what we mean), reliability, and competence. These are the signals of trust that we naturally look for, so it is important that leaders emphasize this practice by demonstrating them. Being proactive in sharing information, being honest, acknowledging the contributions of others, and following through on commitments all help build trust.
3. Extending Trust
   The leader often needs to demonstrate trust first. This requires an act of courage. If we expect trust, we must show trust. People trust what they see. Demonstrating trusting and trustworthy behaviors sends a powerful signal.

Overcoming Fear of Conflict

Building trust is the foundation and requirement for being able to engage in conversations that matter. Conflict is about the productive exchange of diverse ideas and opinions in a focused efficient and unfiltered way so that it can improve both the team's decision-making and the relationships among its members.

But even after trust has been established, teams may still have difficulty with conflict. One reason is that people's viewpoint on and comfort levels with conflict differ radically. Some people are comfortable using compelling communication and arguing passionately; others hesitate to air the mildest of dissenting opinions because they don't want to offend anyone.

Leaders have an important role in building a team's capability to engage in productive conflict. One of the toughest challenges a leader encounters in fostering constructive dissent is the instinct to shield team members from harm. This often leads to prematurely halting disagreements, which inadvertently denies individuals the opportunity to build conflict resolution skills or even to get to the bottom of a dissenting opinion. It is akin to parents who overprotect their children from sibling quarrels – while well-intentioned, this approach can weaken relationships by depriving people of chances to grow through managing disputes. It may also leave unresolved tensions lingering beneath the surface.

For this reason, leaders need to exercise restraint and allow conflicts to be aired and resolved organically, and even to encourage them. Many leaders struggle with this because they feel a loss of control equates to failure in their role. However, a leader's ability to model healthy conflict behavior and conflict resolution is key. When leaders avoid necessary and constructive conflicts, they inadvertently nurture dysfunctional patterns within the team.

Conflict should always be about an issue. Unfortunately, the reasons we can't resolve complex issues often have nothing to do with the issues themselves, but rather with any number of unrelated but important obstacles that team members may find hard to discuss. It is up to the team leader to create conditions where differences can be explored.

Analyzing Patterns of Conflict

Not all conflicts are constructive however, and it is important that leaders can recognize and mitigate patterns of conflict that are less about addressing an issue constructively and more about drama. In the book *Conflict without Casualties*, Regier (2017) suggests many of us shy away from engaging in conflict because, he suggests, we are being invited/coerced into taking up a role in the "drama triangle". Extrapolating from the transactional analysis movement of the 1970s, Regier suggests that when conflict is used negatively, people are taking up roles in a drama triangle, which perpetuates unhealthy behavior. The three roles are described in Table 10.1, as persecutor, victim, and rescuer.

When conflict is being played out negatively by those engaged in taking up these roles, the persecutors, victims, or rescuers are easy to spot if you know their characteristics.

Next time you are either observing or engaged in a drama conflict, see if you can identify which of the roles the protagonists are taking up, and importantly, what role are they (most likely subconsciously) inviting you to take up? The next discussion about accountability and commitment provides some strategies to avoid falling into taking up one of the drama triangle roles.

Developing Accountability, Commitment, and Attention to Results

According to Patrick Lencioni, developing accountability is about removing ambiguity about assumptions and having the utmost clarity about roles and responsibilities. Gaining commitment follows when there is clarity around standards and role expectations, and members are accountable to themselves and others to these standards. Commitment and accountability are, therefore, reciprocally related and when they are in place attention to results can follow.

Regier (2017, p. 66) writes about the link between managing conflicts without casualties and accountability that is sourced from compassion. He notes that

> Compassion without accountability gets you nowhere. Accountability without compassion gets you alienated. Blending the two is the essence of leadership.

TABLE 10.1
Drama Roles with Associated Beliefs, Behaviors, Justifications, and Projected Emotions

Role/*Belief*	Behavior	Justification	Projected Emotions/ *Myths*
Persecutor/ *I'm okay, you're not okay*	Verbally attacking others; blaming; manipulation; giving ultimatums; coercion	Because you are the problem, it's okay for me to use fear, intimidation, and guilt to get what I want.	Rage, frustration, vengefulness, righteous indignation, arrogance/ *"I can make you feel bad emotionally"*
Victim/ *I'm not okay, you're okay*	Over accommodating; avoiding conflict, lacking in assertiveness, whining, self-doubt	I am the problem, and I hurt because I don't deserve any better. I can get what I want when people feel sorry for me	Sadness, insignificance, fear/ *"You can make me feel bad emotionally"* [or] *"You can make me feel good emotionally"*
Rescuer/ *I'm okay and you would be okay if you accepted my help and were grateful*	Meddling with unsolicited advice; wanting to be needed; withdrawing support to "toughen others up"	I can help because I know best and you would be better off if you accepted my help. I can get what I want by showing you how to be okay	Superiority aloofness/ *"I can make you feel good emotionally"*

Source: Regier (2017, pp. 19–20).

In another book about compassionate accountability (Regier, 2023), he proposes that one thing that makes leaders ineffective is the mistaken belief that team members aren't really up to the tasks set. He says that if we fully believe that team members can fulfill their responsibilities, then they can be held accountable. He calls this "struggling with" someone to assist them to achieve the goals and standards. If you fully believe someone should be able to fulfill their responsibilities then the struggle is with them to support them to achieve by providing open and honest feedback. Sometimes conflict occurs when individuals are not clear on what they should be doing, or to what standards, they don't have high expectations they can be achieved.

Regier (2017) proposes we can break out of taking up a conflict drama-triangle role and move toward accountability (and we would add commitment) by changing the way we behave when responding to others. These include being open with the person and empathizing with them, validating their experiences and okay-ness as a human being; disclosing what you are feeling and experiencing in a way that is honest, transparent, and builds rapport.

Three critical competencies for negotiating creative conflict in support of compassionate accountability include the three skills of openness, resourcefulness, and persistence supported by nine strategies, included in Tool 10.1.

TOOL 10.1 COMPETENCIES FOR NEGOTIATING CONFLICT CREATIVELY AND WITH COMPASSIONATE ACCOUNTABILITY

1. Openness
    - **Empathize** – this is about emotional resonance.
        - Ask questions about what's important.
        - Affirm it's okay to share feelings.
        - Take time to listen, avoid inserting your own interpretations or calling attention to yourself.
    - **Validate** – this is about affirming the other's experience not necessarily condoning their behavior or agreeing with it but sending a message that the experience is real for them.
        - Affirm the importance or impact of what the other has said.
        - Avoid disagreeing discounting or re-interpreting.
        - Thank them for sharing.
    - **Disclose** – this is about sharing how you are doing with what you are experiencing. It allows you to demonstrate trust in yourself.
        - Express what you are feeling being confident in your okayness to express these in healthy ways.
        - Own your disclosure, don't blame it on others.
        - Focus on *your* truth, not *the* truth. If you are talking about anyone other than yourself, you are not being open.
2. Resourcefulness
    - **Gather ideas and opinions**.
        - Show a spirit of inquiry and curiosity.
        - Ask open-ended questions.
        - Generate lots of options.
        - Play out scenarios.
        - Explain the gap between what you want and what you are experiencing.
        - Disagree while respecting other's intentions and dignity.
    - **Build on successes**.
        - Think outside the box.
        - Celebrate small things.
        - Set stretch goals.
    - **Leverage personal strengths**.
        - Explore passion and interests.
        - Focus on strategies instead of inherent qualities.
        - Ask about and affirm intentions.
3. Persistence – this is the discipline of holding self and others accountable and respecting the human dignity of all involved.
    - **State your own boundaries and commitments**.
        - Keep the list small and focused.
        - Be proactive.
        - Be explicit.

- **Reinforce your non-negotiables.**
  - Finish what you start.
  - Own the pain and the glory.
  - Don't give up.
  - Remind others about commitments or goals.
  - Be straightforward about what you want.
  - Use consequences carefully.
- **Accept responsibility and make it right.**
  - Identify the behavior in question.
  - Apologize if you are in the wrong and offer to make it right.
  - Be constructive.

The key message in the final element in Tool 10.1 is that "*you and I are accountable*". Using consequences carefully, for example, is not about setting people up to get hurt but not protecting them from learning how their behavior affects the world around them. Ensure these are non-emotional because any consequence that is serving you to feel justified or to punish is drama.

Overcoming Inattention to Results

To strive for collective purpose, leaders need to work hard to create open communication climates where team members will speak up and share their different perspectives. Tool 10.2 comes from the work of Daniel Coyle (2018) on what leaders can do to build the right kind of culture in their teams. The indicators serve as a tool to self-assess your leadership approach.

The indicators include:

Create belonging

1. Over-communicate your listening. A steady stream of [authentic] affirmations encourages speakers to keep going and builds commitment.
2. Spotlight your own fallibility. Especially early on – "I could be wrong here", "What am I missing?". Actively invite input.
3. Overdo thank yous. There is strong scientific support that doing so ignites cooperative effort and commitment.
4. Preview connection between now and the future. Make small but telling connections between now and a vision for the future, including the role of the other and its achievement.

Create safety

5. Embrace the "bad news" messenger. Let them know how much you value their feedback. They will feel safe to be honest in the future.
6. Create safe collision-rich spaces. Design opportunities for people to mingle and interact.
7. Make sure everyone has a voice. Value full-group contributions.
8. Signal your points of vulnerability. Being honest about what you might have screwed up assists others to do so.

Enable cooperation

9. Over-communicate your expectations. Don't presume cooperation will naturally happen. It needs to be actively encouraged and coached.
10. In conversation, resist the temptation to reflexively add value. Skilled listeners do not interrupt with easy solutions: "here's an idea" or "here's what worked for me". Have the willpower to not interrupt.
11. Use flash mentoring. This is like traditional mentoring but only lasts for a short period of time (e.g., an hour). These brief interactions help break down barriers inside a group and build relationships.
12. As a leader, occasionally disappear. This is about leaving the group alone for sometimes just a few minutes. They are empowered to figure out what they need to do at key moments

Establish purpose

13. Name and rank your priorities. Listing your priorities means wrestling with choices and also includes building and sustaining the team itself.
14. Be 10 times clearer on priorities than you think you should be. Leaders are inherently biased to presume everyone sees the things that they do when in fact they don't.
15. Embrace the use of catchphrases. Successful groups have an internal language that may sound corny but helps with purpose and belonging.
16. Focus on bar-setting behaviors. Spotlight the key task that helps the group move toward achieving a longer sometimes more abstract goal.

TOOL 10.2 ASSESS YOUR LEADERSHIP APPROACH

Self-assessment		Never	Hardly ever	Sometimes	Frequently	Not sure
1	Over-communicate your listening	O	O	O	O	O
2	Spotlight your own fallibility	O	O	O	O	O
3	Overdo thank yous	O	O	O	O	O
4	Preview connection between now and the future	O	O	O	O	O
5	Embrace the "bad news" messenger	O	O	O	O	O
6	Create safe collision rich spaces	O	O	O	O	O
7	Make sure everyone has a voice	O	O	O	O	O
8	Signal your points of vulnerability	O	O	O	O	O

Self-assessment		Never	Hardly ever	Sometimes	Frequently	Not sure
9	Over-communicate expectations	○	○	○	○	○
10	In conversation, resist the temptation to reflexively add value	○	○	○	○	○
11	Use flash mentoring	○	○	○	○	○
12	As a leader, occasionally disappear	○	○	○	○	○
13	Name and rank your priorities	○	○	○	○	○
14	Be 10 times clearer on priorities	○	○	○	○	○
15	Embrace the use of catchphrases	○	○	○	○	○
16	Focus on bar-setting behaviors	○	○	○	○	○

When using Tool 10.2 score your answers as follows:

- Frequently = 1 point
- All other responses = 0 points

Your Score: ☐

Source: Coyle (2018).

MAIN POINTS

- Leaders play a pivotal role in creating an environment that encourages open communication and values diverse perspectives.
- Practicing openness and attention to performance standards helps build trust among team members.
- Fostering a culture of constructive conflict within teams is vital for enhancing decision-making and strengthening dialogue.
- Leaders should avoid shielding their teams from conflict, as this can impede the development of conflict resolution skills and does not support building trust.
- Cultivating an environment where differences of opinion are constructively critiqued promotes growth and improved collaboration.
- Modelling constructive behaviors as a leader is essential for encouraging healthy conflict engagement and in moving toward compassionate accountability.
- Prioritizing belonging, purpose, and clarity contributes to building a resilient team culture.

REFERENCES

Coyle, D. (2018). *The culture code: The secrets of highly successful groups.* Bantam.

Edmondson, A. (1999). Psychological safety and learning behavior in work teams. *Administrative Science Quarterly, 44*, 350–383.

Edmondson, A. (2004). Psychological safety, trust, and learning in organizations: A group-level lens. In R. M. Kramer, & K. S. Cook (Eds.), *Trust and distrust in organizations: Dilemmas and approaches* (pp. 239–272). Russell Sage Foundation.

Edmondson, A. C., Bohmer, R. M., & Pisano, G. P. (2001). Disrupted routines: Team learning and new technology implementation in hospitals. *Administrative Science Quarterly, 46*(4), 685–716. https://doi.org/10.2307/3094828

Lencioni, P. (2006). *The five dysfunctions of a team.* John Wiley & Sons.

Regier, N. (2017). *Conflict without casualties.* Berrett-Koehler Publishers.

Regier, N. (2023). *Compassionate accountability: How leaders build connection and get results.* Berrett-Koehler Publishers.

Warren, A. (2017). *Building swift trust.* Retrieved March 5, 2025, from https://www.influence3.com.au/blog/building-swift-trust

11 Leading with Influence

INTRODUCTION

Decision-making and leadership when under pressure require persuasive and extensive communication for appropriate problem recognition and problem-solving with other stakeholders. This relies on relationship building and engagement with a wide range of stakeholders and building cohesion toward shared mission and vision (Waugh & Streib, 2006). Decision-makers need to be able to adapt, especially in complex VUCA-related conditions, and to accept that addressing conflicting priorities rather than minimizing them will be needed for sustainable change under conditions of uncertainty. This requires proactive stakeholder engagement where negotiation and conflict resolution may be needed to expand and unify disparate constituencies.

STAKEHOLDER ENGAGEMENT

Leading with influence involves recognizing that each stakeholder operates from a unique frame of reference, or mental model (Senge, 1990), shaped by their values, experiences, goals, and priorities. The aim is to comprehend and synthesize these diverse perspectives into a unified understanding that reflects the current situation while considering the interests, potential contributions, and concerns of each stakeholder. This approach also seeks to identify both gaps and commonalities in their activities.

ENHANCING INFLUENCE

According to Ken Blanchard et al. (1985), the key to successful leadership today most often relies on influence, not authority. Influence is the power and the ability to personally affect others' actions, decisions, opinions, or thinking. Ultimately, influence allows a leader to get things done and achieve desired outcomes. Influence is a persuasive process.

It is important to highlight that the objective is to influence others rather than manipulate them. This process is gradual and relies on establishing relationships, recognizing others' perspectives, and having a clear understanding of both your own goals and those of others. Decision-makers should approach each interaction as an opportunity for potential influence, both now and in the future. Influence is inherently linked to trust and reputation. By focusing on these aspects, you can cultivate a foundation of trust that enhances your ability to influence effectively. Glass (2018) notes that when people feel heard, they're more open to hearing differences in perspective.

DOI: 10.1201/9781003566007-13

BUILDING SKILLS IN INFLUENCE

The Centre for Creative Leadership (n.d.) has outlined four ways to strengthen the ability to influence others. They suggest that without the expertise and ability to influence others, the truly important things in work and in life can't be achieved. The four skills decision-makers need to develop include developing a keen sense of organizational intelligence; being able to profile and promote the outcomes of the team; engaging in trust-building and leveraging networks (see also Chapter 12).

ORGANIZATIONAL INTELLIGENCE

Developing organizational intelligence is more than simply understanding formal power and hierarchical relationships. It is about recognizing the reality of politics and who informally has influential power. This requires astute observation while remaining as neutral as possible and ethically advancing organizational goals. The skills of asking the right questions and listening attentively are core skills of organizational intelligence. Being politically savvy also means picking up on nonverbal cues and finding ways to make appeals that are in the collective interests and common good.

TEAM PROMOTION

Influential leaders can cut through the noise to ensure that the work of the team and of the leader is recognized and acknowledged. In doing so, effective leaders can find ways to gather an audience and increase influence for the work being undertaken.

TRUST-BUILDING

As Chapters 9 and 10 highlighted, trust is an important foundational skill, not just to cultivate within teams but also in support of trusted relationships with other stakeholders. This requires leaders to be open and vulnerable themselves. There is always a balancing act between pushing into areas of discomfort to expose challenges and support. As Regier (2023) notes, effective leaders are tough-minded on standards and tender-hearted with people. This requires listening carefully to concerns and leading with empathy, as discussed in Chapter 10.

When looking to influence others, it is important to remember the context. Effective and ethical leaders adapt their approaches to fit different situations, carefully selecting when and how to exert their influence. Leaders must have a clear understanding of their values, goals, and perspectives when implementing their influence strategies. This authenticity enhances the effectiveness of their efforts. To prepare for leading with influence, the Centre for Creative Leadership suggests engaging in self-reflection by considering the following questions:

- Who am I trying to influence?
- What is the context? How much support will I need from this person?
- Why is their support for my idea important?

- What strategies can I employ, and how can I build rapport with them?
- What reactions do I expect to receive?
- What mutual points of agreement can I use?
- How can I end on a positive note no matter what the outcome is?

By reflecting on these questions, leaders can strategically navigate the complexities of influencing others and build meaningful connections and relationships.

Asking the Right Questions and Active Listening

The skills of asking the right questions and listening are critical to leading with influence.

Asking the Right Questions

The ability to ask thoughtful, probing questions is essential, for understanding the other person's perspective. When you understand where people are coming from, you know what information will resonate with them. You can frame your question in a compelling way because you've taken the time to understand what's important to them. Asking insightful questions has the added benefit of making people feel valued and listened to. When people feel heard, their defensive walls break down and they're more open to what you have to say.

Consider the following questions:

- Illustration questions: "Can you share or describe what this plan would look like if everything went perfectly?"
- Clarification questions: "You just said that credibility was very important to you. What, exactly, does credibility mean to you?"
- Consequence questions: "If this problem persists, what other issues might it cause?"

Listening for Better Understanding

One of the most critical skills in the leadership toolbox is listening for understanding (see also Chapter 6). Listening is vitally important; sadly undertaught; physically and mentally taxing; and in the aftermath of a crisis is even more difficult. According to Abrahams and Groysberg (2021), listening has three aspects:

- **Cognitive** – paying attention to all the information, both explicit and implicit, that you are receiving from the other person, comprehending it, and integrating it.
- **Emotional** – staying calm and compassionate during the conversation, including managing any emotional reactions (e.g., annoyance, boredom) you might experience.
- **Behavioral** – conveying interest and comprehension verbally and nonverbally. This is important not just for kindness, if people do not feel listened to, they will stop sharing information.

Abrahams and Groysberg (2021) offer nine suggestions for active listening outlined below.

1. **Repeat People's Last Few Words:** Research shows that repeating the last few words of what someone has said can be an effective way to show you are listening. It makes the speaker feel heard, keeps you engaged in the conversation, and gives both parties a moment to process their thoughts.
2. **Avoid Rephrasing Unless Necessary:** Instead of rephrasing what the other person has said, which can lead to confusion and emotional friction, only do so when you need to confirm your understanding. Clearly state, "I'm going to paraphrase to ensure I understand correctly".
3. **Use Nonverbal Cues Naturally:** While eye contact, attentive posture, and nodding can signal that you are listening, forcing these behaviors can be counterproductive. If nodding or prolonged eye contact feels unnatural, mention at the beginning of the conversation that you might be more non-reactive and ask for their understanding.
4. **Pay Attention to Nonverbal Signals:** Active listening involves observing both verbal and nonverbal cues. Tone of voice, facial expressions, and body language often convey emotions and motivations beyond the spoken words.
5. **Ask Questions:** Inquiry is key to ensuring comprehension and enhancing the other person's feeling of being listened to. Asking questions fosters clarity and prompts the speaker to elaborate on important points. Ask more questions than you think you need to.
6. **Minimize Distractions:** Strive to reduce external noise and interruptions, as well as internal distractions. If you're preoccupied, take a moment to refocus before the conversation begins, especially if you anticipate emotional topics.
7. **Acknowledge Limitations:** If you know you might struggle to listen attentively due to fatigue or unfamiliarity with the conversation topic, communicate this upfront. If you lose track during the discussion, admit it and ask the person to clarify their points.
8. **Avoid Rehearsing Responses:** Resist the urge to prepare your reply while the other person is speaking. Take a moment to reflect after they finish, allowing yourself to fully absorb what they've said before formulating your response.
9. **Monitor Your Emotions:** If you find yourself having an emotional reaction, slow the conversation down. Practice deeper repetition and focus on your breathing to maintain composure. This helps prevent knee-jerk reactions that could disengage the other person or lead to defensiveness.

By incorporating these strategies, leaders can foster better communication, enhance mutual understanding, and strengthen relationships through effective listening.

How to Recognize When You're Being Influenced

As a leader who is accustomed to being the influencer, it can be challenging to recognize when the dynamics shift, and others are trying to persuade you. Here are some steps to help you identify when you may be under the influence of others:

1. **Cultivate Self-Awareness:** Start by reflecting on your own biases and perspectives. Understanding your strengths and areas for development will help you recognize when you might be swayed by others. Consider questions like: "What triggers my typical responses, and how might these impact my ability to see other viewpoints?"
2. **Recognize Your Reactions:** Pay attention to your emotional and cognitive responses during discussions. If you find yourself feeling defensive, uncertain, or overly agreeable, it may indicate that you're being influenced. Notice any shifts in your thoughts or feelings that don't align with your initial stance.
3. **Assess the Context:** Evaluate the setting and dynamics of the conversation. Are there power imbalances at play? Is the person trying to influence you using persuasive language or emotional appeals? Being mindful of these elements can help you assess the influence being exerted.
4. **Engage in Reflective Thinking:** Ask yourself what it would take for you to be more aware of others' influence. Think critically about your typical approaches in influence situations and consider how you can balance them to be more effective with others.
5. **Use Collaborative Consultation:** If you notice you are being influenced, consider leveraging a consultative approach. Share your thoughts, desires, and suggestions in an open discussion. This fosters collaboration and can be particularly valuable if you find yourself at odds with others. Negotiating from a place of mutual understanding can lead to better outcomes.
6. **Apply Your Influencing Skills:** Reflect on the strategies you've honed for influencing others. These same skills can help you navigate times when you're being influenced. By remaining aware and engaged, you can ensure that interactions remain productive and aligned with your values and objectives.

By developing these skills and increasing your self-awareness, you'll be better equipped to identify when you're being influenced and respond in a way that maintains your integrity and effectiveness as a leader.

THE ART OF NARRATIVES AND STORYTELLING

Throughout history humans have connected and learned through storylines. Bauer (2021) suggests that stories help us pay attention – particularly in this contemporary era, when vying for people's focus is more coveted than ever.

According to Bauer, when we see or hear a story, the neurons in our brain fire in the same patterns as the speaker's, a process known as "neural coupling", also referred to as "mirroring".

Research reported in Bauer (2021) shows that in story telling these processes occur across many different areas of the brain and can induce a shared contextual model of the situation. Several areas of the brain fire up when listening to stories, including the motor and sensory cortices and frontal cortex.

Leaders can use the power of storytelling to enhance influence.

- Stories provide useful moments for reflection.
- Stories can help others emotionally connect.
- Stories can provide clear "take home" messages that are useful in energizing commitment.

In this respect, stories facilitate strategies to help overcome barriers to moving toward purposeful action. Instead of doubt and fear, they can help support urgency, hope, and confidence (Ganz, 2011). This includes enabling others to connect to existing values and beliefs that help alleviate barriers to action.

STRATEGIES FOR PRACTITIONERS – USING STORYLINES

In crises, leaders also need to recognize the power of storylines when others have developed their own which are not a good fit with sense-making about what might be unfolding.

A challenge for leaders is how to shape storylines so that partners, stakeholders, and the community receive the intended message and can thus shape their own explanations of what is happening and what they need to do. Members of the community, for example, start forming a storyline to explain an event very quickly to settle their own cognitive dissonance. The storyline will be based on the information available to them through their own experiences, friends, social media and official sources.

If leaders want a set storyline that supports their common goals, then the creation of that storyline must happen quickly and be reinforced through as many means as possible. For example, the Deepwater Horizon event was changed from the largest oil spill ever to the biggest response to an oil spill ever (Walsh, 2011): very different messages. In disaster management, we have changed the storyline from disaster "victims" to disaster "survivors". This has had a very big impact on how people perceive their situation and act. What are the storylines you need to promote? Leaders need to strategically think about what the key storylines could be. In a crisis, plans to build and reinforce key storylines are essential (Tool 11.1).

Remember that influencing others is not about winning; it is fundamentally about collaboratively addressing and solving problems. While it is natural to seek your own objectives, it is important to consider the long-term implications of your influence on relationships and outcomes

TOOL 11.1 CRAFTING NARRATIVES WITH INFLUENCE

Effective influence hinges on persuasive communication, and there are various frameworks to help convey your message clearly, credibly, and compellingly. Glass (2018) recommends a strategic approach to expanding influence by considering the responses to the following questions. When crafting requests, it is important to factor in the thoughts and feelings of others.

1. **Describe the Current State:** Outline what is currently happening. Present a clear picture of the existing situation.
2. **Establish a Need/Problem:** Identify the issues with the current state. Why is there a need for change? Make it clear why addressing this problem is essential.
3. **Describe the Desired State:** Illustrate what the ideal outcome looks like. Emphasize the importance of making this change or following the suggested course of action.
4. **Satisfy or Solve:** Specify what actions need to be taken. What steps should be implemented to address the problem or move forward? It's helpful to frame this from a mutual perspective or align with their viewpoint.
5. **Visualize:** Paint a picture of the benefits that will arise once the problem is resolved. Why is this outcome better than the current situation?
6. **Call to Action:** Request commitment or agreement on the proposed steps forward.

MAIN POINTS

- In VUCA-related conditions, leaders need to address issues collaboratively, through active listening, resolving conflicts, and fostering cohesion among diverse stakeholders.
- Leaders can utilize verbal and nonverbal cues, ask questions for clarity, minimize distractions, acknowledge limitations, and avoid rehearsing responses during conversations to enhance influence and understanding.
- Storyline Formation: Communities quickly create storylines to explain events, influenced by personal experiences, social media, and official sources. Leaders must ensure these storylines are established and reinforced effectively.
- Strategic Storyline Promotion: Decision-makers under pressure can increase their influence by identifying and promoting key storylines to shape community understanding and response.

REFERENCES

Abrahams, R., & Groysberg, B. (2021). *How to become a better listener.* Retrieved May 24, 2025, from https://hbr.org/2021/12/how-to-become-a-better-listener

Bauer, T. (2021). *The neuroscience of storytelling.* Retrieved May 25, 2025, from https://neuroleadership.com/your-brain-at-work/the-neuroscience-of-storytelling/

Blanchard, K., Zigarmi, P., & Zigarmi, D. (1985). *Leadership and the one-minute manager: Increasing effectiveness through situational leadership. William Morrow and Company. Inc.*

Centre for Creative Leadership (n.d.). *Four Keys to strengthen your ability to influence others.* Retrieved June 6, 2019, from www.ccl.org/articles/leading-effectively-articles/4-keys-strengthen-ability-influence-others

Ganz, M. (2011). Public narrative, collective action, and power. In *Accountability through public opinion: From inertia to public action* (pp. 273–290). The World Bank.

Glass, A. (2018, October 5). *Five principles to follow if you want to influence others.* Forbes Coaches Council. Retrieved May 27, 2025, from https://www.forbes.com/councils/forbescoachescouncil/2018/10/05/five-principles-to-follow-if-you-want-to-influence-others/

Regier, P. D. (2023). *Compassionate accountability: How leaders build connection and get results.* Berrett-Koehler Publishers.

Senge, P. (1990). *The fifth discipline: The art and practice of the learning organization.* Doubleday.

Walsh, B. (2011, April 20). "One year later": Thad Allen, hero of the Gulf, reflects on the BP oil spill. *Time Magazine.* Retrieved May 29, 2025, from https://time.com/archive/6917276/one-year-later-thad-allen-hero-of-the-gulf-reflects-on-the-bp-oil-spill/

Waugh, W. L. Jr, & Streib, G. (2006). Collaboration and leadership for effective emergency management. *Public Administration Review, 66,* 131–140. https://doi.org/10.1111/j.1540-6210.2006.00673.x

12 Leveraging Networks and Alliances

INTRODUCTION

Leaders with expertise in influencing others recognize and cultivate the power of networks. Organizations are increasingly dynamic; they morph in size and shape over time. Influential leaders recognize that their personal networks must also be dynamic, and they continually grow and strengthen their networks. They're also strategic about choosing how and when to tap into this network.

Pfiefer (2013) discussed the importance of reaching out to diverse stakeholders and connecting to them in loosely coupled alliances to facilitate shared situation awareness, particularly important in VUCA-related conditions. The skills of situation awareness involve the capacity to search and make sense of information about the threat environment. Networking supports this when leaders connect to a diverse group of people and sources of information at and away from the problem-solving environment to develop a comprehensive picture of the situation.

Networking involves the ability to anticipate and conceive ways to customize and improvise, coordinating with others to achieve what cannot be achieved alone (Pfiefer, 2013). Doing so requires skills in negotiation, collaboration, and partnering which involves the ability to recognize interdependencies and to do so with others in adaptive ways.

NETWORKS AS SOCIAL CAPITAL

According to the Kellogg School of Management (Uzzi, 2006), effective networks are rich in social capital characterized with trust, diversity, and brokerage. Trust, Diversity, and Brokerage are regarded as the three pillars upon which an effective network structure is built.

Trust

Trust is about demonstrating both personal integrity and effectiveness (i.e., there is trust that this person can deliver). However, it is also easier to trust those who are like us, and so this can be an impediment to diversity.

Diversity

Uzzi (2006) observed that those who were most successful seemed to fully transcend this barrier and work with those who were sometimes drastically different (i.e., demonstrating diversity). Networking also complements existing skills sets by connecting to others with diverse backgrounds.

DOI: 10.1201/9781003566007-14

Brokerage

Brokerage is about linking up with people because of the links they have and doing so in a way that aids in mutual reciprocity. Therefore, for brokerage to be useful in building networks and alliances, it needs to be active and employed on a regular basis, which in turn aids trust.

For successful leaders, networking is a critical approach to acquire information, gain access to key influencers, and access to resources. Networking enables leaders to keep a pulse on what might be happening in different sectors, calling on various experts for guidance and advice. Networking also enables access to sources of informal power through building relationships.

LEVERAGING NETWORKS

Effective leaders have diverse and open networks that they strategically use and grow. In their guidebook on leadership networking, Grayson and Baldwin (2011) propose the following six elements are useful in building effective networks.

1. **Authenticity:** Be sincere in how you share information in an authentic way.
2. **Resource sharing:** Effective leaders engage in reciprocity.
3. **Use power thoughtfully:** Power comes from building your reputation and your alliances.
4. **Skillful communications:** This involves being able to clearly articulate your needs and what you can offer to accomplish goals.
5. **Effectively negotiate:** This is about having an astute sense of what is important and where the boundaries are.
6. **Manage conflict:** Part of doing this is about having an awareness of the perspective of others, looking for points of mutual agreement, and being constructive in attempting to resolve points of conflict.

ADAPTIVE LEADERSHIP AND NETWORKING

To leverage networks and build alliances, leaders need to be able to adapt without compromising their own values and ethics base. Research undertaken by Calarco (2020) at the Centre for Creative Leadership suggests leaders need three types of flexibility to adapt to pressures involved in building networks and alliances during crises. An inability to adapt is seen as the hallmark of many failures. The three types of flexibility needed are cognitive, emotional, and dispositional.

Cognitive Flexibility

Cognitive flexibility is the ability to use different thinking strategies and mental frameworks. Leaders who have cognitive flexibility can incorporate different thinking strategies and mental frameworks into their planning, decision-making, and management of day-to-day work. They can simultaneously hold multiple scenarios in mind and can see when to shift and inject a change. Cognitive flexibility indicates nimble, divergent thinking, an interest in developing new approaches, the ability to see and leverage new connections, and the propensity to work well across the organization. These leaders readily learn from experience and recognize when old approaches don't work.

Emotional Flexibility

Emotional flexibility involves the ability to vary your approach to dealing with emotions and those of others. Leaders with emotional flexibility vary their approach to dealing with their own and others' emotions – an area that many leaders often fail to consider. An emotionally flexible leader is comfortable with the process of transition, including grieving, complaining, and resistance. Adapting to change requires give and take between the leader and those experiencing the change. A leader without emotional flexibility is dismissive of others' concerns and emotions and shuts down discussion. At the same time, an emotionally adaptive leader moves the change or agenda forward.

Dispositional Flexibility

Dispositional flexibility is the ability to remain optimistic and, at the same time, realistic. Leaders who display dispositional flexibility operate from a place of optimism grounded in realism and openness. They will acknowledge a bad situation but simultaneously visualize a better future. They are neither blindly positive nor pessimistic and defeatist. Ambiguity is well-tolerated. Leaders with dispositional flexibility see change as an opportunity rather than as a threat or danger.

By learning and practicing behaviors that boost your cognitive, emotional, and dispositional flexibility, you can become more adaptable and, in turn, help others to adapt. Calarco (2020) proposes five things that leaders can do to be adaptable:

- Be curious. Ask lots of questions. Wonder, explore, and consider before you judge and decide.
- Don't get too attached to a single plan or strategy. Have Plan B (and C) at the ready.
- Create support systems. Don't go it alone. Look to mentors, friends, coaches, trusted peers, professional colleagues, family members, and others to serve as your support system in times of change.
- Understand your own reaction to change. You must be clear about your own emotions and thoughts about changes, so you can be straightforward with others.
- Immerse yourself in new environments and situations. Do this when you are confronted by change – but get practice by joining activities, meeting new people, and trying new things on a regular basis.

BUILDING ALLIANCES THROUGH CROSS-SILO ENGAGEMENT

According to Edmondson et al. (2019), leaders of teams can help encourage their team members to build their own alliances and thus aid the situation awareness of the team by engaging in the following strategies.

Develop and Deploy Cultural Brokers

In most teams, there are people who already excel at interface collaboration. They usually have experiences and relationships that span multiple sectors, functions, or domains and informally serve as links between them. Edmondson et al. (2019) call these people cultural brokers. Cultural brokers promote cross-boundary work in one of two ways: by acting as a "bridge" or as an "adhesive".

Bridges

A bridge offers themselves as a go-between, allowing people in different functions or geographies to collaborate with minimal disruption to their day-to-day routine. Bridges are most effective when they have considerable knowledge of both sides and can figure out what each one needs.

Adhesives

Adhesives, in contrast, bring people together and help build mutual understanding and lasting relationships. Adhesives facilitate collaboration by vouching for people and helping them decipher one another's language. Unlike bridges, adhesives develop others' capacity to work across a boundary in the future without their assistance.

Encourage People to Ask the Right Questions

It is nearly impossible to work across boundaries without asking a lot of questions. Enquiry is critical because what we see and take for granted on one side of an interface is not the same as what people experience on the other side. Ask the question:

- Who are our most valued stakeholders?

and once identified,

- Which relationships are most important for creating value for them?

See also the material in Chapter 11 for asking effective questions.

Getting People to See the World Through Others' Eyes

Assisting others to engage in mindful listening (Chapter 6) and asking the right questions will help. Leaders of teams can also become role models in showing interest in what others are seeing and thinking by asking questions. According to Edmondson et al. (2019), doing so has a stunning effect. It prompts team members to do the same.

If possible, leaders of teams should look to organize cross-silo dialogues. The goal is to get everyone to share knowledge and work on synthesizing that diverse input into new solutions.

Broaden Your Team Member's Vision

This can occur when leaders teach their team members the art of enquiry. Doing so invigorates a sense of curiosity (see also creating a sense-making mindset in Chapter 7). Questions and inquiry are the secret to productive work relationships – but they must be driven by genuine interest in understanding another's view. Leaders can also bring people from diverse groups together on initiatives.

STRATEGIES FOR PRACTITIONERS

Some practice strategies for developing networking capabilities, according to Uzzi (2006), include:

1. **Be Passionate About Your Work:** Demonstrate a strong track record by consistently delivering results. Cultivating a robust personal brand that reflects your passion will attract like-minded individuals who bring energy and support, enhancing your professional journey.
2. **Build Trust as a Foundation:** Trust is essential for effective networking and is established through reliable and consistent behavior over time. While ideas and tangible outcomes are important, relationships take precedence. Focus on nurturing long-term relationships, as the benefits will naturally follow.
3. **Embrace Diversity:** Step outside your comfort zone to connect with individuals who hold different perspectives and opinions. Engaging with a diverse network of people who can challenge your ideas will foster personal and professional growth. Maintain relationships with industry leaders, former professors, and other subject matter experts who can offer valuable insights.
4. **Regularly Assess Your Network:** Periodically evaluate your network to identify gaps or areas for improvement. Take proactive steps to diversify your connections by adding individuals from various backgrounds. Consider utilizing the shared activity principle – collaborating with others on significant projects – to explore opportunities for connection.
5. **Reciprocate to Foster Mutual Success:** Recognize and reach out to individuals in your network that you may have overlooked. Provide assistance to those who need it, thereby optimizing the flow of information and resources within your network. Their achievements will contribute to your own success, reinforcing the idea that networking is a two-way street.

The Role of Network Governance in Managing VUCA Events

One way to cope with complexity in dynamic conditions is through network governance. Cross-sector collaboration in responding to VUCA events is seen as a form of network governance. That is when autonomous partners engage in addressing a common issue or problem (Pilemalm & Mojir, 2023). In this respect, there is horizontal collaboration between organizations with shared interests, leading to collective service delivery or decision-making without any clear top-down governing mechanism.

Network governance is primarily described as a phenomenon referring to horizontal collaboration between autonomous actors with shared interests, leading to collective service delivery or decision-making. Its core assumption is that the network consists of autonomous actors who interact to make policies and perform service delivery in a horizontal pattern without any clear top-down governing mechanism. This results in sometimes different organizations managing a VUCA-related event being temporarily collocated, sharing resources, or mutually agreeing to divvying up tasks.

However, the convergence of multiple organizations to respond and manage a VUCA-related event is not without challenges. Boersma et al. (2021) highlight some of the power dynamics that can be in play as various actors attempt to claim and negotiate their interests and responsibilities which may hinder the desired coordinated approach. Practitioners would be wise to be mindful of the strategies to name up and manage micro-politics, as discussed in Chapter 7 and to consider some of the practices of demonstrating political astuteness (discussed in Chapter 18).

Conduct a Ripple Map to Identify Needed Alliances

In thinking about who you might need to develop collaborative relationships to resolve the crisis, it is useful to consider the ripples or consequences a crisis will have on various stakeholders and entities.

Analyze the Ripple Effects of a VUCA Event

To analyze ripple effects in disaster coordination, Holley (2020) talks about the importance of understanding ripple effects from events such as disasters.

STEP 1. Develop a ripple map of consequences for your event.

STEP 2. Identify groups/organizations/communities of interest who are involved or impacted by the event. In this step, it is important to ask yourself "*so what?*" (what was the consequence) and then "*so who?*" (who does that impact). Ripples are then mapped since one ripple (first-order consequence) may lead to a second ripple (second-order consequence) and so on.

STEP 3. Draw the ripple effects of the consequences naming up what and who are impacted. Now you have a consequence network diagram for the groups/organizations/collectives you have listed and for whom the event will have consequences (see Figure 12.1).

Considering the consequences of an event and its impacts on others is a way of moving into anticipatory thinking and understanding who the event will impact or involve. You can use this map to reach out to those likely to be affected, and if you do so ahead of time, you can begin to build a more coordinated effort and assist those stakeholder organizations to appreciate what might be coming their way that they may not yet appreciate. Doing so will support those and is likely to gain support for your goals and agendas to resolve or manage the issues at hand.

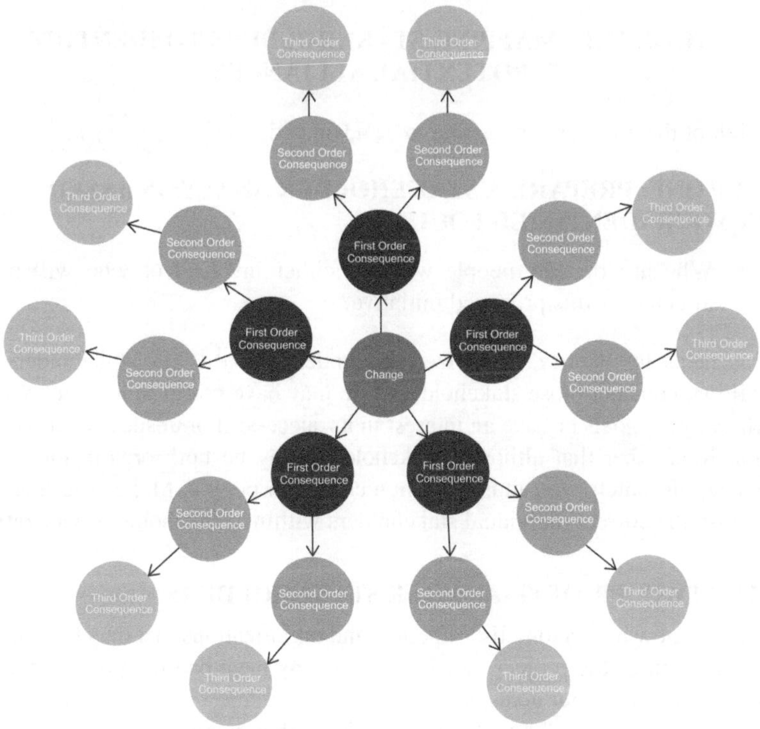

FIGURE 12.1 A ripple map identifying consequences of change.

A consequence or ripple map also allows you to consider the stakeholders involved and to analyze where you will need to give energy and attention.

Conduct a Stakeholder Analysis

One way to analyze a network of potential alliances is to conduct a stakeholder analysis. Stakeholder maps are important visual models of the key people and relationships that impact on your role. Do not make the mistake of thinking that hierarchy or traditional lines of authority are always the routes by which decisions are made. Take the time to uncover the underlying relationships and influence that key stakeholders have. With this insight, you can tap into the real sources of power and persuasion. While this is something that people do intuitively in small projects, it is something that leaders need to do actively for larger problem-solving endeavors under VUCA conditions.

Develop a Map of Stakeholder Influence

Listed below are steps to construct a stakeholder influence map. Think about a recent event you were involved with (or one in the future where there are likely to be differing stakeholders and interests). These stakeholders might be people internal to the organization, the sector, or in other areas of community (Tool 12.1).

TOOL 12.1 MAPPING STAKEHOLDERS TO IDENTIFY POTENTIAL ALLIANCES

Think of the aim, goal or agenda to be advanced.

STEP ONE: PREPARE A STAKEHOLDER ANALYSIS BASED ON YOUR IDENTIFIED FOCUS

- Who are the key people who are either involved or who will be affected by this proposed initiative?

This helps to identify, prioritize, and understand the key stakeholders. In addition, consider those stakeholders who may have power in assisting you to achieve your goals or have an interest in its successful or unsuccessful conclusion. Remember that although stakeholders may be both organizations and people, ultimately, you must communicate with people. Make sure that you identify the correct individual stakeholders within a stakeholder organization.

STEP TWO: PRIORITIZE YOUR STAKEHOLDERS

You may now have a long list of people and organizations that may be affected by the event or the project. Some of these may have had the power either to block or advance your goals. Some may have an interest in what you are doing, others may not care about what you are doing that much.

STEP THREE: MAP YOUR STAKEHOLDERS ON THE POWER/ INTEREST GRID

Map out your stakeholders on a power/interest grid (see Figure 12.2) and classify them by their power over your work and your goals, and by their interest. For example, the person you are reporting to is likely to have high power and influence over whether you can fulfil your role and achieve your goals. Your family might have had high interest in what you are doing, but unlikely to have had power over it. Someone's position on the grid shows you the actions you need to ensure you take with them:

- High power and interested people – these are the people you must fully engage and make the greatest efforts to satisfy.
- High power less interested people – put enough work in with these people to keep them satisfied, but not so much that they become bored with your message.
- Low power interested people – keep these people adequately informed and talk to them to ensure that no major issues are arising. These people can often be very helpful with the detail of your role.
- Low power less interested people – again, monitor these people, but do not bore them with excessive communication.

STEP FOUR: ASSESS FOR EXISTING NETWORKS TO IDENTIFY FUTURE ALLIANCES

Study the map and consider your existing relationships with your stakeholders.

Form a stakeholder engagement plan based on the influence map you have created that will allow you to communicate with and hopefully build strong alliances with these stakeholders.

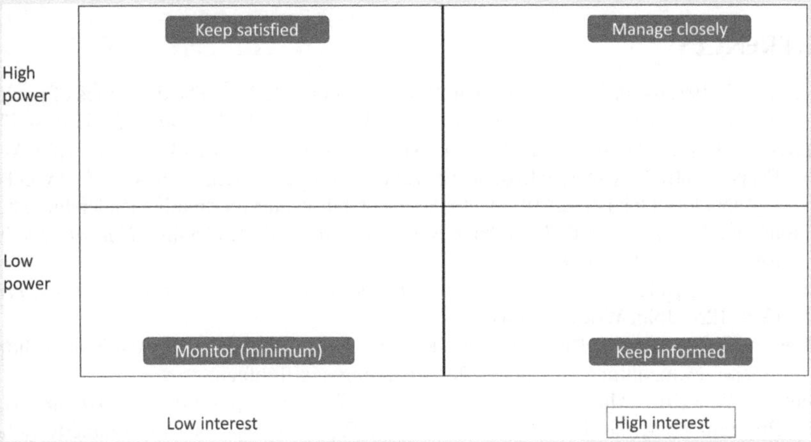

FIGURE 12.2 Mapping stakeholders for cultivating alliances. (From Thompson, R. (2017). *Stakeholder analysis: Winning support for your projects.* Retrieved May 4, 2025, from www.mindtools.com/pages/article/newPPM_07.htm)

MAIN POINTS

- Leaders working in VUCA recognize the need for dynamic networks that grow and strengthen over time, mirroring the changing landscape of their organizations and their environments.
- Key Elements of Effective Networks include Trust, Diversity, and Brokerage, and these work to facilitate mutual reciprocity and social capital.
- For successful leaders, networking is crucial for acquiring information, accessing key influencers, and leveraging resources, enabling them to stay informed about various sectors and access informal power.
- According to Grayson and Baldwin (2011), there are six elements for building effective networks. These are Authenticity; Resource Sharing; Thoughtful Use of Power; Skillful Communications; Effective Negotiation; and Conflict Management
- Leaders can use adaptive strategies that can support collaboration and partnering without compromising their values and ethics. These include developing cognitive, emotional, and dispositional flexibility.

- Strategies for practitioners developing networking capabilities are outlined which include being passionate and delivering consistent results; building foundational trust; embracing diversity of thinking, engaging in regularly assessing your network, and reciprocating to foster mutual success.
- A tool is offered which provides an opportunity for practitioners to assess the stakeholders implicated in achieving their goals and to identify gaps in their existing network with the purpose of identifying future potential alliances.

REFERENCES

Boersma, K., Ferguson, J., Groenewegen, P., & Wolbers, J. (2021). The dynamics of power in disaster response networks. *Risk, Hazards & Crisis in Public Policy, 12*(4), 418–433.

Calarco, A. (2020). *Adaptable leadership: What it takes to be a quick-change artist.* White Paper, Centre for Creative Leadership. Retrieved June 1, 2025, from https://www.ccl.org/wp-content/uploads/2021/01/adaptable-leadership-center-for-creative-leadership.pdf.

Edmondson, A. C., Casciaro, T., & Jang, S. (2019). Cross-silo leadership. *Harvard Business Review, 97*(3), 130–139.

Grayson, C., & Baldwin, D. (2011). *Leadership networking: Connect, collaborate, create* (Vol. 125). John Wiley & Sons.

Holley, J. (2020). Ripple effect mapping. Retrieved 20th August 2025. Available at https://juneholley.medium.com/ripple-effect-mapping-a4c014bfae88

Pfiefer, J. W. (2013, March). *Crisis leadership: The art of adapting to extreme events.* Program on Crisis Leadership Discussion Paper Series. Harvard Kennedy School. Retrieved June 2, 2025, from https://www.hks.harvard.edu/sites/default/files/centers/research-initiatives/crisisleadership/files/Pfeifer%20Crisis%20Leadership–March%2020%202013.pdf.

Pilemalm, S., & Mojir, K. Y. (2023). Digitalized cross-sector collaboration for an effective emergency response: Emerging forms of network governance. In H. J. Scholl, E. E. Holdeman, & F.K. Boersma (Eds.), *Disaster management and information technology: Professional response and recovery management in the age of disasters* (pp. 269–305). Springer International Publishing. https://doi.org/10.1007/978-3-031-20939-0_13

Thompson, R. (2017). *Stakeholder analysis: Winning support for your projects.* Retrieved May 4, 2025, from www.mindtools.com/pages/article/newPPM_07.htm

Uzzi, B. (2006, December 1). *Leadership network analysis.* Retrieved June 2, 2025, from https://www.kellogg.northwestern.edu/faculty/uzzi/ftp/teaching%20materials/Example_std_papers/emp70_Paper.pdf

Part 3

The Situation

When practitioners, their teams, and other stakeholders are impacted by volatility, uncertainty, complexity, and ambiguity, the situation is dynamic and this influences the ways in which those involved frame their thinking. Part 3 explores the ways in which practitioners need to address VUCA-related environments, recognizing the various forces of change, only some of which are in the practitioner's control.

Chapter 14 considers the ways in which VUCA-related environments can flex and change, leading to a range of possible outcomes. Under these circumstances, decision-makers need to be mindful of weak signals indicating change and anticipate potential disruptive points. After considering various ways to think about uncertainty and risk, strategies to identify disruptive and culminating points are discussed. A tool to assess trigger points to develop various courses of action is provided.

Chapter 15 extends this discussion by taking up the differences between predictions, forecasts, and foresight, making a case for the mindset needed to engage in anticipatory thinking which supports agility. Strategies for practitioners to apply foresight are discussed. The chapter also discusses how learning and adaptation is required in the moment. That is, in dynamic VUCA-related situations, failure will be part and parcel of the work being undertaken and using failure, for example to learn and adapt becomes an important attribute. A tool supporting scenario thinking is outlined that draws on conducting a pre-mortem as part of contingency planning.

The following three chapters each explore the implications for practitioners engaging with others impacted by situations underpinned by VUCA conditions. Chapter 16 looks at the likely psycho-social impacts of a VUCA event on human behavior in communities. It discusses how trauma is collectively experienced and provides some strategies practitioners can use when engaging with affected community members. Chapter 17 extends this discussion by considering the paths practitioners must navigate when working with both order and chaos. That is, practitioners still need

DOI: 10.1201/9781003566007-15

to uphold legislation and organizational standards and policies and at the same time make space for communities to work through processes of grief and to be able to take control of community-led decision-making. This means practitioners are not in charge of these decisions and so a different mindset is needed, one of hosting conversations. The next chapter takes up the ways in which practitioners need to be mindful of social license, that is, the conditions by which their actions are supported by others through legitimacy, credibility, and trust. Self-awareness for the practitioner is increased when they are mindful of the ways in which their actions are interpreted by others and when they can operate with political astuteness.

13 Leadership and Organizational Crises

INTRODUCTION

An organizational crisis is a situation that poses a significant threat to the organization's survival, distinguishing it from an emergency event that may be either anticipated or unanticipated but requires coordinated resource deployment. While managing an emergency focuses on immediate response, managing an organizational crisis is more strategically oriented, typically involving high-level officials and emphasizing long-term impacts (McNulty & Marcus, 2020).

ORIGINS OF AN ORGANIZATIONAL CRISIS

Organizational crises can emerge from various sources, including:

- **Pandemics:** Public health crises can disrupt operations and threaten organizational viability (Boin et al., 2021).
- **War:** Geopolitical conflicts can have far-reaching consequences for organizations, particularly those in affected regions (Opatska et al., 2024).
- **Climate Change:** Environmental factors can lead to crises through natural disasters, regulatory changes, and social pressures (House of Lords, 2021).
- **Malicious Attacks:** Cyber-attacks targeting critical infrastructure can cripple operations and erode stakeholder trust (Hajkowicz, 2016).
- **Accidents:** Unforeseen accidents can lead to significant organizational crises, prompting urgent responses (Alexander, 2012; Burns, 2014).

Stakeholder Interpretation of Crises

An event may escalate to a crisis based largely on how stakeholders perceive and interpret the situation. Some common interpretations that may contribute to the perception of an organizational crisis include:

- **Negligence:** The belief that the organization failed to meet its obligations.
- **Lack of Compassion:** Perceived indifference to stakeholder concerns or suffering.
- **Corruption:** The suspicion of unethical practices within the organization.

DOI: 10.1201/9781003566007-16

- **Loss of Trust:** Erosion of confidence among stakeholders, making recovery challenging.
- **Criminal Activity:** Allegations or evidence of illegal behavior that damages the organization's reputation.
- **Inadequate Response:** Perceptions that the organization has failed to act effectively or in a timely manner.
- **Poor Moral and Ethical Standards:** Beliefs that the organization lacks strong moral principles.
- **Service Disruption:** Inefficient management leading to disruptions that impact service delivery.
- **Dishonesty and Greed:** Perceived motives of self-interest overshadowing ethical responsibilities.
- **Lack of Commitment:** Inability to demonstrate a commitment to stakeholders during crises.

An organizational crisis often arises when a triggering event coincides with public or stakeholder outrage. This outrage can amplify the crisis, making it imperative for leaders to navigate the situation effectively and restore confidence.

NAVIGATING ORGANIZATIONAL CRISES AND THEIR SURVIVAL

Understanding the different types of organizational crises is helpful for effective leadership during these challenging situations. Collins et al. (2023) draw on the crisis typology developed by Coombs and Holladay (1996) to explore how various crisis types impact crisis leadership. This typology categorizes crises based on stakeholders' perceptions of their origins and intentions, allowing leaders to tailor their responses effectively.

CRISIS TYPOLOGY OVERVIEW

In a literature review conducted by Collins et al. (2023), organizational crisis events were classified according to two main dimensions: whether they are internal or external to the organization, and whether they are intentional or unintentional. This categorization influences how stakeholders interpret the crisis and sets the tone for the behaviors expected of leadership. Below is a summary of the typology and associated leadership metaphors (Table 13.1).

Crisis Responses by Type

External and Unintentional Crises: Leaders as Shepherds

In crises emanating from external sources and perceived as unintentional (e.g., natural disasters like Hurricane Katrina), leaders are often viewed as shepherds (Collins et al., 2023). Their role involves protecting and guiding both the organization and those affected through the crisis. Leaders focus on preparatory activities, mitigating damage, and supporting recovery efforts. They centralize decision-making to ensure swift

TABLE 13.1
Typology of Crisis Leadership Origins and Behavior

Origin and Attribution	Example	Leadership Metaphor	Key Behaviors
External and Unintentional	Hurricane Katrina	Leaders as Shepherds	Centralizing decision-making authority; preparing and planning; mitigating damage; facilitating learning.
External and Intentional	Terrorism (e.g., 9/11, Oslo bombing)	Leaders as Saints	Empowering others; emphasizing a collective vision; providing support; using compelling rhetoric.
Internal and Unintentional	Product defects; contamination of AIDS medication	Leaders as Spokespersons	Managing impressions; public apologies; disclosing failures; accepting responsibility.
Internal and Intentional	Enron scandal	Leaders as Sinners	Attributing atonement; emphasizing restorative actions; focusing on ethical conduct.

Source: Collins et al. (2023).

and effective responses, utilizing command-and-control approaches to facilitate quick adaptations. Key leadership behaviors include:

- **Identifying Early Warning Signs:** Recognizing potential crises before they escalate.
- **Facilitating Recovery:** Guiding stakeholders toward recovery while promoting learning from the experience.
- **Providing Compelling Communication:** Making sense of the situation and maintaining credibility by communicating effectively.
- **Demonstrating Empathy:** Exhibiting confidence and compassion to foster a collective commitment to action.

External and Intentional Crises: Leaders as Saints

In crises that are externally caused by intentional acts (e.g., terrorism), leaders are often viewed as saints who provide guidance and support to help navigate fear and uncertainty (Collins et al., 2023). Successful leaders in these situations create a unified vision, helping stakeholders make sense of their circumstances. Their capabilities include:

- **Creating a Sense of Us vs. Them:** Identifying a common enemy to galvanize collective action.
- **Offering Support and Hope:** Providing comfort while managing negative emotions such as fear and anger.
- **High Emotional Intelligence:** Demonstrating emotional understanding and compassion to foster positive stakeholder interactions.

Internal and Unintentional Crises: Leaders as Spokespersons

When crises arise from internal factors but are unintentional (e.g., product defects), leaders are held accountable and must work to restore stakeholder trust (Collins et al., 2023). Effective leadership behaviors include:

- **Honesty and Transparency:** Communicating openly about failures and taking responsibility for them.
- **Public Apologies:** Acknowledging mistakes and expressing regret to minimize reputational damage.
- **Taking Accountability:** Implementing measures to prevent future failures and offering compensation to affected parties.

Internal and Intentional Crises: Leaders as Sinners

Crises stemming from internal, intentional actions (e.g., scandals like the Enron case) require leaders to address ethical breaches. In these situations, stakeholders often view leaders as sinners in need of atonement (Collins et al., 2023). Effective leadership strategies encompass:

- **Public Atonement:** Leaders must publicly apologize for their actions or those of their organization.
- **Rebuilding Trust:** A commitment to reform and ensuring ethical practices moving forward.
- **Internal Stakeholder Support:** Recognizing that employees may feel stigmatized and addressing their concerns while promoting healing.

Navigating organizational crises requires a clear understanding of the crisis types and the appropriate leadership responses that align with stakeholders' perceptions. By adopting the appropriate behaviors connected with each crisis type, leaders may enhance their effectiveness in managing the crisis and work towards restoring trust and organizational integrity, drawing on appropriate crisis management governance.

CRISIS MANAGEMENT GOVERNANCE

In an organizational crisis, it is not just the organizational leader who has a role to play but also the crisis management team. The crisis management team plays a crucial role in coordinating with emergency management teams that deploy resources, as well as collaborating with crisis communications teams. Additionally, depending on the type of organization, a board of directors may have specific responsibilities and obligations during a crisis.

Decision-makers under pressure work with their crisis management teams to take decisive actions to manage the organizational crisis effectively. As outlined in Table 13.2, these actions include maintaining situational awareness and ensuring that strategic imperatives are upheld throughout the crisis.

TABLE 13.2
Actions for a Crisis Management Team

Focus	Conformance Actions	Performance Actions
External	**Providing accountability to stakeholders including regulators** Who needs to know about the situation and who will do it? Government, Regulator, Donors, Partners?	**Strategic thinking and direction setting** What are the implications of this situation for the future direction of the organization? What are the threats? What are the opportunities? New partners?
Internal	**Monitoring and supervising** What changes to the organization's control system might need to be made considering this situation?	**Policy making and guidance to the organization** What policy and guidance do the CEO and executive team need to give managers?

Source: Second author (Parsons).

Roles of the Crisis Management Team

The crisis management team is responsible for several critical activities, which include:

- **Accountability to Stakeholders:** Ensuring transparency and accountability in actions taken during the crisis.
- **Monitoring and Supervising:** Overseeing the execution of crisis response plans and ensuring alignment with organizational goals.
- **Policy Making and Guidance:** Establishing policies that guide organizational responses and behavior during the crisis.
- **Strategic Thinking and Direction Setting:** Providing long-term vision and goals to navigate the crisis effectively.

The crisis management team must ensure that external perceptions of the organization's crisis management efforts reflect a commitment to several key principles (Tool 13.1):

- **Prioritizing Well-being:** Demonstrating a caring and compassionate approach towards people affected by the crisis.
- **Responsibility:** Taking appropriate responsibility for the crisis and its management.
- **Timely Stakeholder Engagement:** Addressing stakeholder needs promptly to maintain trust.
- **Honesty and Integrity:** Making decisions based on legal and ethical guidelines, ensuring transparency.
- **Visibility and Communication:** Being available, visible, and maintaining open lines of communication with stakeholders.
- **Cooperation with Investigations:** Willingly cooperating with independent investigations to enhance transparency and accountability.

TOOL 13.1 KEEPING A STRATEGIC FOCUS

To maintain a strategic focus, the crisis management team should receive regular briefings from the incident management team regarding the technical response. The following are questions that decision-makers can engage their crisis management team in addressing:

- **Assessing Gaps:** Is there a gap between organizational actions and stakeholder expectations?
- **Evaluating Stakeholder Confidence:** How is the event affecting stakeholder confidence in the organization?
- **Impact on Reputation:** What potential impact will the event have on the organization's reputation?
- **Policy Adjustments:** What urgent changes are needed in the organization's policies, processes, and structures?
- **Understanding Stakeholder Expectations:** How will this event shift stakeholder expectations of the organization?
- **Community Reaction:** What kind of reaction can be anticipated from the community—will there be outrage?
- **Regulatory Changes:** What potential regulatory or legislative changes could arise from this event?
- **Assessing Risks:** Are there any legal or financial risks to consider?
- **Maintaining Capacity:** What actions need to be taken to ensure the organization can respond effectively moving forward?
- **Preparing for Inquiries:** Will there be an inquiry or investigation, and what preparatory actions should begin now?

STRATEGIES FOR PRACTITIONERS

While crises inherently carry risks of blindness and failure (Kapucu & Van Wart, 2008), they can also present opportunities to enhance credibility and maintain a social license to operate (O'Reilly, 2020; see also Chapter 18). Despite the challenges posed by adversity (Rosenthal et al., 2001), crises can open pathways to improvement and a better future (Johnson & Murray, 2021).

MAIN POINTS

- Organizational crises can arise from both internal and externally sourced events and can be based on intentional or unintentional acts, necessitating adept leadership approaches.
- Public Accountability: Leaders must provide accountability to stakeholders, addressing both the implications of the crisis and the necessary changes to the organization's control systems.

- Addressing Ethical Breaches: In cases of internal, intentional crises, leaders must publicly apologize and commit to reforms to rebuild trust and ensure ethical practices moving forward.
- Leaders must identify early warning signs and facilitate recovery while maintaining stakeholder confidence and credibility.
- Crises present both risks and opportunities, potentially threatening reputation and operational capacity but also offering chances for growth and improvement.
- Understanding stakeholder expectations and preparing for potential regulatory changes is essential for effective organizational crisis response.
- Internal Stakeholder Support: Recognizing and addressing the concerns of employees during a crisis is crucial for promoting healing and maintaining organizational integrity.

REFERENCES

Alexander, D. E. (2012). The 'Titanic Syndrome': Risk and crisis management on the Costa Concordia. *Journal of Homeland Security and Emergency Management*, *9*(1), 1–25.

Boin, A., McConnell, A., & 't Hart, P. (2021). *Governing the pandemic: The politics of navigating a mega-crisis*. Springer Nature.

Burns, W. (2014). Off course: The baffling case of MH370 and the crisis management that survived it. *Corporate Public Affairs*, *24*(1), 4.

Collins, M. D., Dasborough, M. T., Gregg, H. R., Xu, C., Deen, C. M., He, Y., & Restubog, S. L. D. (2023). Traversing the storm: An interdisciplinary review of crisis leadership. *The Leadership Quarterly*, *34*(1), 101661.

Coombs, W. T., & Holladay, S. J. (1996). Communication and attributions in a crisis: An experimental study in crisis communication. *Journal of Public Relations Research*, *8*(4), 279–295.

Hajkowicz, S. (2016). *Australia 2030: Navigating our uncertain future*. CSIRO.

House of Lords (2021). *Preparing for extreme risks: Building a resilient society. Select committee on risk assessment and risk planning* (Report of Session 2021–22, Issue Dec). www.parliament.uk; https://publications.parliament.uk/pa/ld5802/ldselect/ldrisk/110/11002.htm

Johnson, E., & Murray, F. (2021). What a crisis teaches us about innovation. *MIT Sloan Management Review*, *62*(2), 7.

Kapucu, N., & Van Wart, M. (2008). Making matters worse: An anatomy of leadership failures in managing catastrophic events. *Administration & Society*, *40*(7), 711–740.

McNulty, E. J., & Marcus, L. (2020). Are you leading through the crisis… or managing the response. *Harvard Business Review*, *68*, 531–532. https://hr.umich.edu/sites/default/files/are_you_leading_through_the_crisis_or_managing_the_response.pdf

O'Reilly, B. (2020, May 3). *Don't let a good crisis go to waste. Instead, use it as a catalyst for innovation*. Singularity Hub. Retrieved May 27, 2025, from https://singularityhub.com/2020/05/03/dont-let-a-good-crisis-go-to-waste-instead-use-it-as-a-catalyst-for-innovation/

Opatska, S., Johansen, W., & Gordon, A. (2024). Business crisis management in wartime: Insights from Ukraine. *Journal of Contingencies and Crisis Management*, *32*(1), e12513. https://doi.org/10.1111/1468-5973.12513

Rosenthal, U., Boin, A., & Comfort, L. K. (2001). *Managing crises: Threats, dilemmas, opportunities*. Charles C Thomas Publisher.

14 Managing Situations of Deep Uncertainty

INTRODUCTION

As noted in the Introduction and Chapter 1, operating in VUCA-related contexts requires decision-makers to have complex sense-making and decision-making skills to plan and respond accurately to VUCA challenges (see Aydiner et al., 2023; Grant, 2023; Warren & Frankel, 2022). In short, this means that in contemporary high-consequence contexts to successfully plan and act in dynamic environments leaders need to think differently.

Part of the challenge in dynamic situations in any context is that practitioners are frequently not good at recognizing weak signals that change is on its way, which includes that existing plans and actions are not working until it is almost too late. This chapter explores the challenges facing practitioners operating in the context of uncertainty, disruption, and change, and the implications for planning in these environments.

To begin, one might question why attempt to plan at all under VUCA conditions? In short, in the words of General Dwight D. Eisenhower

> In preparing for battle, I have always found that plans are useless, but planning is indispensable.

This is because planning provides practitioners with:

- A strategy to manage an event.
- A desired goal or set of goals to be achieved.
- A way of reducing (though not eliminating) uncertainty.
- Identification of actions to achieve a particular result or outcome.
- A roadmap to measure and monitor team performance.

However, as the quote from Eisenhower suggests, planning is situated within an environment where there are many forces, including those that cannot be controlled. These forces will inevitably impact on the parameters and concerns to be addressed and the scope of managing an event.

CLASSIFYING UNCERTAINTY

Uncertainty is unavoidable when thinking about VUCA situations. Because knowledge of the present is typically incomplete, decision-makers can't be sure what will happen in the future. They can reduce uncertainty due to lack of knowledge by obtaining more information or asking experts for advice. However, they cannot eliminate uncertainty entirely, so we must find ways to describe and communicate it.

DOI: 10.1201/9781003566007-17

The idea of "known knowns, known unknowns and unknown unknowns" was popularly attributed to Donald Rumsfeld. In terms of classifying certainty, "*known knowns*" are the things we know that we know. For example, in tossing a coin, we know we cannot predict the outcome since, with a fair coin toss, the odds are 50/50. In addition, there are "*known unknowns*". These are elements that we know we do not know because of the complexity involved (e.g., is the coin a fair one?). "*Unknown unknowns*" are future events or circumstances that cannot be predicted. Only the first two (known knowns and known unknowns) can be reduced.

Uncertainty and Risk

Making decisions using uncertain information can be uncomfortable – decisions would be far easier if we knew what the consequences of each choice would be – but this is an unavoidable position. It is also important to note that there is a difference between uncertainty and risk. Risk typically refers to the likelihood of a future unplanned event or unintended consequence that can be assigned a numeric probability.

The *Uncertainty Toolkit for Analysts in Government* (GAD, 2020) notes that when dealing with dynamic or high-consequence decisions, best estimates are generally not good enough to make an informed decision. For example, Option A may appear better than Option B when looking at the most likely outcome of each. But if the uncertainty in A is greater than in B, it might carry an unacceptable chance of much worse outcomes (GAD, 2020).

Taking account of uncertainty – and being seen to do so – is important for public trust (see also Chapter 18). The consequences of decisions are always uncertain to some degree. For each option, a range of outcomes is possible. Implying certainty about a particular outcome can damage public trust when things turn out differently.

So, what does this mean for attempting to plan in situations with VUCA conditions?

FRAMING FUTURES

We often frame a plan to manage a situation as if the actions leading to an outcome will be a straight line – that the outcome will either be a preferred outcome, business as usual or a worst-case scenario. The Futures Cone in Figure 14.1 illustrates the spectrum of potential futures.

In Figure 14.1, the center line and circle signify the expectation that the future will be just like the present or "business as usual". Widening out the circle is both the "plausible" future, that is with current knowledge, what could happen and even be a preferable future – that is what we want to happen. It is often the case that practitioners selectively filter information to support their preferred outcome and ignore any signals to the contrary. However, as Figure 14.1 suggests, there are additional potentialities that result in other things that could happen (possible future), even if in the present moment those futures seem preposterous. This means that the forces acting in the situation are likely to be more dynamic than a linear outcome might suggest. In fact, in dynamic environments, the outcomes may be highly uncertain.

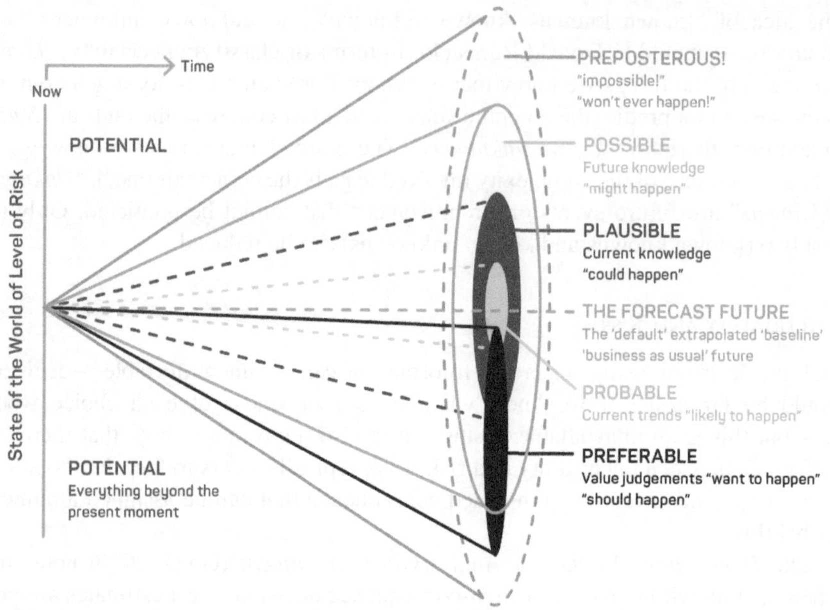

FIGURE 14.1 The futures cone. (From Australian Government, Department of Home Affairs (2019). *Climate and Disaster Risk: What they are, why they matter and how to consider them in decision making: Guidance on Scenarios* (p. 10). Commonwealth of Australia. Retrieved June 2, 2025, from www.aidr.org.au/media/6932/04-scenarios.pdf.)

Uncertainty and Vortices

Instead of a straight line between now and a planned pathway to a desired future outcome, it is useful to consider the potential trajectories and the forces for change likely to be operating in the situation. In this respect, the beginning of a planning pathway may look more like the cone of uncertainty in Figure 14.2. The cone of uncertainty can be used to consider a range of possible futures – to show how there is no one version of the future but multiple possibilities – see Figure 14.2. It is important to explore and make sense of this range of future possibilities using scenario thinking. Doing so helps overcome the enemies of planning (biases and other blind spots—see Chapter 3) and to avoid the ways they derail plans. This means taking account of the wider set of drivers (or forces of change) that are likely to be in play.

Disruption

Like weak signals, the impact of disruptive events when interpreting and managing VUCA events typically does not emerge out of the blue. It is important not just to look for the disruptions that are obvious (i.e., will impact the center of core business) but also to look to those at – or potentially moving in from – the margins. It is vital to pick signals of disruptive change across an entire ecosystem

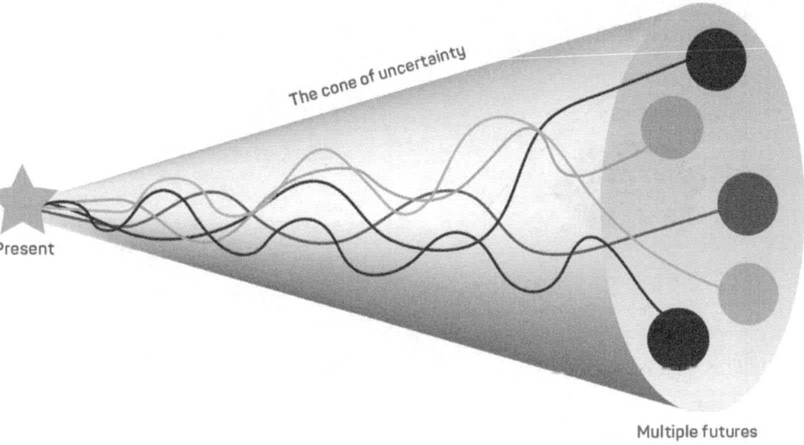

FIGURE 14.2 The cone of uncertain possibilities.

of stakeholders, regulators, suppliers and more since these will ultimately impact on what practitioners can and cannot achieve. Doing so minimizes the chances of being caught by surprise.

Agility comes from recognizing signals and their potential for disruptive impact. This means overcoming blind spots or denial, which, if not addressed will result in an escalating crisis that could have been mitigated if weak signals had been identified and addressed earlier. Certain continuities, such as climate change and advances in technology, for example, can become discontinuities if developments reach a disruption threshold – e.g., if extreme weather and coastal erosion lead to mass migration, or if advances in artificial intelligence lead to widespread employment dislocations (Hoehn et al., 2018, p. 11).

HORIZON SCANNING

Horizon scanning is the process of looking for early warning signs of change in a dynamic situation. Horizon scanning tries to detect the weak signals of change that are embedded in background noise, which may become part of a significant pattern if connected with other information or viewed through a different lens. Identifying weak signals supports anticipatory thinking because once identified as a possibility identification can help a faster response to emerging issues. In so doing, horizon scanning assists in preparing for possible disruptive events. In line with the drivers or forces influencing the uncertainty is a recognition of the differences in planning for different horizons. Figure 14.3 illustrates how key issues change over time (Government Office for Science, 2024).

The present and near future are encapsulated in what is referred to as Horizon 1 (H1) in Figure 14.3. H1 encompasses the issues that are currently significant, visible, and well understood, serving as the primary concerns that decision-makers and stakeholders are actively addressing. However, it is important to recognize that the

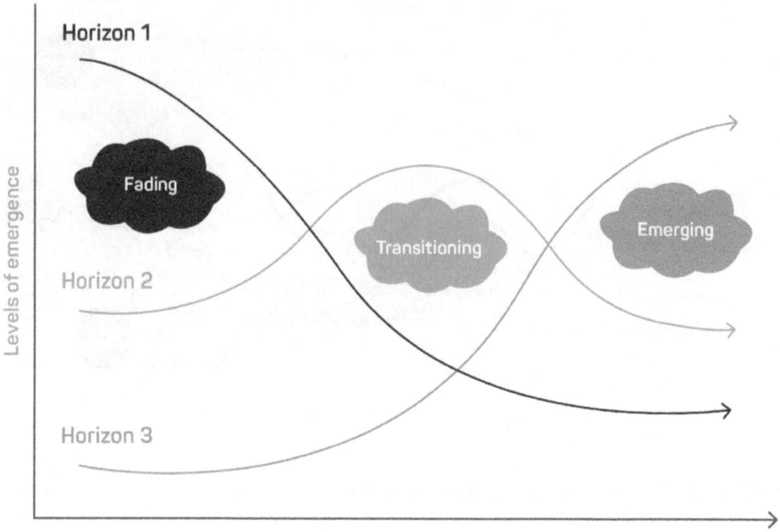

FIGURE 14.3 The three-horizon model. (From Government Office for Science. (2024). *The futures toolkit: Tools for futures thinking and foresight across UK government.* Retrieved June 2, 2025, from https://www.gov.uk/government/publications/futures-toolkit-for-policy-makers-and-analysts/the-futures-toolkit-html)

relevance of H1 issues may diminish over time, particularly in complex and rapidly changing situations. H1 issues can be supplanted by emerging matters or events categorized under Horizon 2 (H2).

Although H2 issues may not yet be fully visible, the underlying driving forces are already at work. Therefore, decision-makers must monitor these potential H2 issues closely, assess possible outcomes, and modify plans proactively to meet future needs. This process ideally involves collaboration with stakeholders to ensure their perspectives and requirements are integrated into planning.

During dynamic situations, new developments may arise that shift focus to Horizon 3 (H3), leading to the need for further action and addressing these evolving priorities. The implications are that practitioners must remain vigilant and adaptable, recognizing that the landscape of pressing issues can transform, necessitating ongoing evaluation and collaborative planning.

Identifying Weak Signals

Understanding the dynamics of VUCA situations in the present and the near future is often straightforward. Key trends and events at Horizon 1 (H1) are prominent, making their impacts clear. However, as practitioners extend their focus further into the future, these signals become less distinct, and it becomes increasingly challenging to identify patterns and interpret their significance.

Horizon scanning is the practice of recognizing and deciphering these weak signals. By nature, such signals lack robust evidence, which can make it difficult for those engaged in scanning to determine their potential impact or articulate why they might consider a weak signal to be significant. This uncertainty, however, should not be a deterrent. In the absence of solid evidence, it is essential for scanners to trust their gut feelings when a weak signal suggests it may hold strategic importance for the future. In considering the middle time horizon, this intuitive approach becomes even more valuable. Thinking creatively, what might be some weak signals that the anticipated prediction is not likely to occur? What other forces or drivers are increasing in their influence and impact on the event?

In summary, in dynamic environments practitioners need to:

- Consider the full range of possible outcomes when using analysis to inform decisions, not just the "best estimate".
- Challenge those providing predictions where information on uncertainty is absent or inadequate.
- Ask questions about how to interpret the uncertainty and its implications.

Avoid getting too caught up with the present, the immediate demands and what seems obvious. It is important to spot patterns of change, emerging trends, potential surprises, and disruptors earlier, so there is more time to respond.

UNCERTAINTY AND MAPPING RISK

Mapping critical uncertainties is also helpful when examining what needs to be considered when planning in dynamic situations. In evaluating uncertainty and levels of risk, Waldron et al. (2020) suggest three questions:

1. What is the nature of the risk associated with the potential future change?
2. What is the likelihood that the risk plays itself out?
3. If the risk played out, how much of an impact would it have?

The implication is that decision-makers need to consider the priorities when mapping critical uncertainties (see Table 14.1).

TABLE 14.1
Implications for Planning from Mapping Critical Uncertainties

Impact/Level of Uncertainty	Low/Known	Moderate	High/Unknown
Critical	Priority – plan first	Address these	Critical game changer
Moderate	Plan for these	Plan for these	Address these – may alter course
Low	Monitor	Monitor	Monitor and reassess

Source: GAD (2020).

In dynamic situations, it is particularly important to ensure that *known unknowns* are articulated to identify possibilities and risks. The next chapter will discuss strategies for forecasting in dynamic and uncertain situations.

STRATEGIES FOR PRACTITIONERS: IDENTIFYING DISRUPTIVE AND CULMINATING POINTS

One way of looking for weak signals is to be alert to disruptive points and culminating points. In the joint Appreciation Process published by the Australian Defence Force (2019, pp. 3–33) a decisive point is described as:

> A significant operational milestone that exists in time and space or the information domain which constitutes a key event, essential task, critical factor or function that, when executed or affected, allows a commander to gain a marked advantage, or contributes to achieving success.

A decisive point includes the following characteristics:

- It is the articulation of a purpose, outcome, task, or effect.
- It is measurable in terms of time, space, and magnitude.
- It is expressed in the past tense.

An example of a disruptive point would be: *the deadline to make an informed decision to evacuate has been reached*. Disruptive points are important to anticipate as they will throw off the intended outcome of plans and lead to a different future state, as illustrated in Figure 14.4.

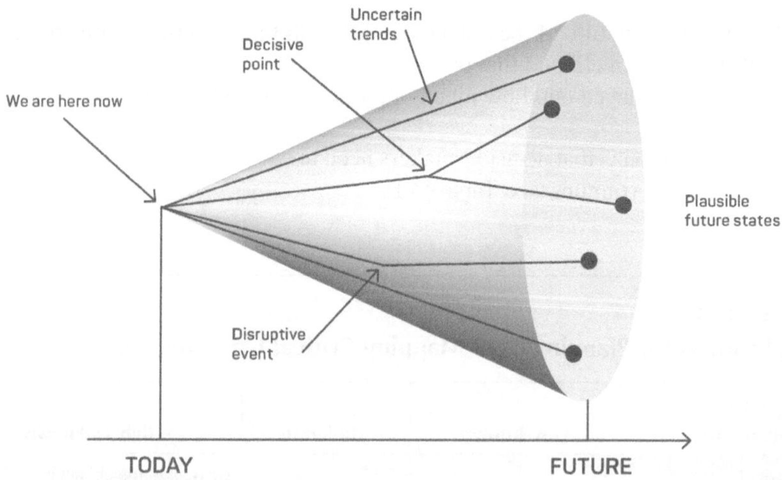

FIGURE 14.4 Links between decisive points and plausible future states. (From Rowland, E. R., Cross, M. S., & Hartmann, H. (2014). *Considering multiple futures: Scenario planning to address uncertainty in natural resource conservation*. US Fish and Wildlife Service. Retrieved July 18th, 2025, from https://www.cakex.org/sites/default/files/documents/Final%20Scenario%20Planning%20Document(2).pdf)

According to the Australian Defence Force Manual (2019, pp. 4–13), a culminating point is the point in time and location where a force will no longer be stronger than the adversary/threat and planners risk losing the initiative. In the context of making decisions under pressure in a high-consequence VUCA event, an example might be: *the point where the number of flood rescues required exceeds the number of flood rescue resources available*. Tool 14.1 is provided to assist practitioners to identify what disruptive points might be in their own contexts and to then scope out what might be the courses of action required once the disruptive or culminating points have been reached. Doing so supports anticipatory thinking.

TOOL 14.1 ASSESS TRIGGER POINTS AND COURSES OF ACTION

STEP 1: DETERMINE KEY TRIGGER POINTS

Think of a situation you are managing. What might be some trigger points that will require certain actions if those trigger points are reached? (For example, not being able to restore power for longer than six hours). Try to identify clear trigger points that will prompt decision-making and action on the part of your team. Use the space below to populate your answers. Your trigger points need to be monitored regularly to help make timely decisions and agile plans in relationship to the circumstances.

STEP 2: CREATE A PORTFOLIO OF ACTIONS

The actions will likely have some resourcing impacts and implications for effort and occur over whatever time horizon the action would be implemented. Actions then need to be considered in terms of their impact on best-, moderate-, and worst-case scenarios (see Chapter 15). The portfolio of trigger points and actions can then be mapped as some will be critical to choosing between different courses of action.

Trigger Points	Courses of Action
Example:	• Request back-up generator supplies
Power has not been restored for longer than six hours	• Consider evacuation
Your own example	*Your own example*

LEARNING ON THE FLY

When operating in complex VUCA contexts, iterative learning and improvisation will play an important part in any operation. While preparedness- training and exercising is important, under VUCA conditions, planning goals need to be subject to continuous review and adjustment.

TABLE 14.2
CYNEFIN Framework

Complex Contexts	**Complicated Contexts**
Probe – sense – respond (i.e., emergent practice)	Sense – analyze – respond (i.e., good practice)
Chaotic Contexts	**Simple Contexts**
Act – sense – respond (i.e., novel practice	Sense – categorize – respond (i.e., best practice)

Source: Snowden and Boone (2007).

David Snowden (Snowden & Boone, 2007) proposed that decision-makers under pressure need to recognize what context they are in. Snowden's Cynefin (meaning different subsystems in which we operate) framework is outlined in Table 14.2. Reading the table starts in the bottom right-hand quadrant and represents contexts where practitioners face a clear and simple problem, and where how to response is well known and based on best practice. Here, practitioners can sense a situation, categorize it as easily recognizable and know how to respond. A situation might, however, be more complicated (top right-hand side) which requires some kind of analysis or expertise. Here a practitioner may call on expertise to help analyze the situation. However, in volatile, non-routine events, we move into a state of complexity (Top left-hand side). Here much is unknown, and practitioners must guide their thinking by trying some things out and seeing what happens, hence the reference to "probe-sense- respond". This is the domain of emergent knowledge as multiple stakeholders, not just known experts, are needed to contribute their knowledge. For example, this was the situation in the Chilean Mine Disaster in 2010, when 33 Chilean miners were trapped underground with no known solution to drill into rock to reach the trapped miners. Multiple stakeholders across the globe were engaged to try and find a solution which ultimately was successful some 70 days after the collapse (Rashid et al., 2013). The bottom left-hand quadrant indicates contexts that are chaotic, where decision-makers must act first to stabilize the situation, then make sense of what is happening then respond. In the framework, there is also a 5th dimension, which sits in the middle and represents confusion or disorder. This can happen when practitioners cannot recognize the context that should guide their actions and respond inappropriately.

Of course, there are also situations where all four quadrants need to be attended to. Snowden and Boone (2007) described how in January 1993, following a gunman murdering seven people in a restaurant in a Chicago suburb, Police Deputy Chief Walter Gasior was faced with managing several situations at once. This included acting quickly to restore order (chaos), responding to media and collectively engaging with other stakeholders as to their needs (complex), seeking advice from experts in relation to trauma counselling for the children who witnessed the attack (complicated) and keep the police department operating on a routine basis (simple). The challenge for practitioners is to recognize which situation they face and to act accordingly.

In exploring the processes of practitioners who excel in times of crisis, Ansell and Boin (2019) propose that these leaders recognize that uncertainty is inherent in all situations of complexity. These leaders are operating closer to what Snowden and Boone (2007) called contexts that are complex and potentially chaotic. Here, the plan comes from acting, seeing what happens and then responding again in an iterative process.

Unlike planning by pre-defined design, this is planning by feedback as the situation unfolds. Under novel and dynamic conditions, a practitioner may have to act their way into understanding their environment. This represents an iterative relationship between action, perception, and understanding that comes together in a process that unfolds in small steps (Ansell & Boin, 2019). In this way, the goals develop through trialing and learning.

PRAGMATISM IN NAVIGATING COMPLEXITY

In writing about Taming Deep Uncertainty, Ansell and Boin (2019) propose that practitioners need to incorporate some of the principles of the philosophy of pragmatism into their planning and decision-making repertoires.

When one is faced without a clear picture of the situation – given the dynamic conditions – it is hard to make critical decisions and coordinate a complex response network; it is also hard to effectively communicate to anxious citizens and concerned stakeholders (Ansell & Boin, 2019).

The principles of pragmatism proposed by Ansell and Boin (2019) include:

1. Acknowledging the need for constant calibration in the face of emerging doubts and new facts. This requires an attitude of inquiry and doubt. Complexity and uncertainty are acknowledged rather than reduced.
2. Avoid framing the problem as a "yes or no" dichotomy because this can blind us to the complexity of the dynamic environment. That is, rather than defining the goals as "now or never" or "do or don't". Instead, look for scales rather than categories, grey zones rather than hard boundaries. Deliberation informs this process.
3. A third critical principle is that of experimentalism. Goals are provisional and subject to continuous revision as new information becomes available. Decision-makers treat their options as suppositions: Once an option is "enacted", practitioners should carefully monitor the impact (or lack thereof) of their action and adapt where necessary. A series of small acts in adaptive fashion will help decision-makers wrestle down the imperatives of uncertainty. Rapid feedback allows for subsequent refinement of the plan. This "probe and learn" approach also helps decision-makers to avoid making irreversible commitments.
4. The fourth principle is to adopt an attitude of "bricolage". Bricolage is about exploiting available resources, combining them in improvisational ways to address specific needs. Existing roles/responsibilities or groups may be deployed for new purposes. It is grounded in the reality of what is possible (vs. what is desirable). It tells decision-makers not to reach for the moon.

According to Ansell and Boin (2019, p. 1090), "In the pragmatist perspective, perfection is the enemy of the good".

These principles emphasize continuous inquiry, a more targeted evaluation of whether an interpretation of the situation continues to line up with changing facts on the ground. Emphasis is given to triangulating between different sources of information to construct coherent interpretations and then testing these interpretations.

> Inquiry, in the Pragmatist sense, calls for a reflexive and deliberative examination of ways to productively address value conflicts. … Decision makers are not just information processors. They must approach situations as creative, moral, and knowledgeable actors.
>
> *(Ansell & Boin, 2019, p. 1090)*

Ansell and Boin also argue that no amount of information can resolve the basic value trade-offs inherent in dynamic environments. Under these conditions, it is important to appreciate that information is incomplete, attention selective, and perspective limited. The dynamic environment is continuously scanned for "cues" and so decision-makers can reflexively consider how cognitive biases and complex system dynamics might hide them.

The pragmatist perspective suggests that when operating in deep uncertainty it is important to maintain a picture of the situation as a hypothesis, continuously exploring different scenarios to cope with uncertainty. Scenario thinking and developing foresight are explored in the next chapter *Foresight for Futures Thinking.*

MAIN POINTS

- Developing the skills of horizon scanning is important to navigate future uncertainties in dynamic situations.
- In doing so, there is a need to recognize weak signals of change, emerging trends, and potential disruptors to respond effectively.
- Decisive points are important to identify because they represent significant operational milestones that will lead to choices of actions and thus different futures.
- Decisive points should articulate a purpose, be measurable in time and space, and be expressed in the past tense.
- Identifying clear trigger points that prompt decision-making actions should be monitored regularly.
- Creating a portfolio of actions provides for anticipatory thinking that considers the impacts on various scenarios (best, moderate, worst) and maps critical actions against trigger points
- The principles of the philosophy of pragmatism are helpful for practitioners to add to their decision-making repertoires. It includes an attitude of inquiry and updating to achieve calibration in the face of emerging information; using scales rather than categories to frame problems; engaging in experimentalism and adopting an attitude of bricolage, which is about exploiting available resources in improvised ways.

REFERENCES

Ansell, C., & Boin, A. (2019). Taming deep uncertainty: The potential of pragmatist principles for understanding and improving strategic crisis management. *Administration and Society*, *5*(7), 1079–1112.

Australian Defence Force (2019). *Procedures. Edition 2 AL3 Joint military appreciation process – plans series.* Retrieved 2nd June, 2025, from https://theforge.defence.gov.au/sites/default/files/adfp_5.0.1_joint_military_appreciation_process_ed2_al3_1.pdf

Australian Government, Department of Home Affairs (2019). *Climate and Disaster Risk: What they are, why they matter and how to consider them in decision making: Guidance on Scenarios.* Commonwealth of Australia. Retrieved June 2, 2025, from www.aidr.org.au/media/6932/04-scenarios.pdf

Aydiner, C., Corbin, T. B., & Tan, C. (2023). Developing critical thinking and effective communication skills in the future aviation workforce. *National Training Aircraft Symposium (NTAS). 37.* Retrieved June 2, 2025, from https://commons.erau.edu/ntas/2022/presentation/37

Government Actuaries Department (GAD). (2020). *Uncertainty toolkit for analysts in government.* Retrieved June 2, 2025, from https://analystsuncertaintytoolkit.github.io/UncertaintyWeb/index.html

Government Office for Science. (2024). *The futures toolkit: Tools for futures thinking and foresight across UK government.* Retrieved June 2, 2025, from https://www.gov.uk/government/publications/futures-toolkit-for-policy-makers-and-analysts/the-futures-toolkit-html

Grant, A. (2023). *Think again: The power of knowing what you don't know.* Penguin.

Hoehn, A., Parasiliti, A., Efron, S., & Strongin, S. (2018, April 27). *Discontinuities and distractions: Rethinking security for the year 2040. Findings from a RAND Corporation Workshop.* RAND Corporation. Santa Monica, CA. Retrieved June 2, 2025, from https://www.rand.org/pubs/conf_proceedings/CF384.html

Rashid, F., Edmondson, A. C., & Leonard, H. B. (2013). Leadership lessons from the Chilean mine rescue. *Harvard Business Review*, *91*(7–8), 113–119.

Rowland, E. R., Cross, M. S., & Hartmann, H. (2014). *Considering multiple futures: Scenario planning to address uncertainty in natural resource conservation.* US Fish and Wildlife Service. Retrieved July 18, 2025, from https://www.cakex.org/sites/default/files/documents/Final%20Scenario%20Planning%20Document(2).pdf

Snowden, D., & Boone, M. (2007, November). A leader's framework for decision making *Harvard Business Review*. pp 1–9.

Waldron, L., Searle, R., & Jaskula-Ranga, A. (2020, May 27). *Making sense of uncertainty: Nonprofit scenario planning in the COVID-19 pandemic.* The Bridgespan Group. Retrieved June 2, 2025, from https://www.abhe.org/wp-content/uploads/2020/05/nonprofit-scenario-planning-COVID-19.pdf

Warren, M., & Frankel, M. (2022, August 11). *Are you thinking clearly? 29 reasons you aren't, and what to do about it.* Coronet.

15 Foresight for Futures Thinking

INTRODUCTION

Foresight is described as "the ability to see through the apparent confusion, to spot developments before they become trends, to see patterns before they emerge, and to grasp the relevant features of social currents that are likely to shape the direction of future events" (Whitehead, 1947, p. 89, cited in Burt & Nair, 2020). This definition accommodates two contrasting understandings of "foresight". In the first, foresight refers to insights gained from anticipations of the future, whereas in the second, foresight is attained through a better understanding of the "present" in the here and now.

Two additional concepts are particularly relevant when operating in VUCA situations. The first is the idea of a *forecast*, which involves making a calculation or estimation of future events based on data from past occurrences, along with recent trends. This approach allows decision-makers to project potential outcomes and prepare accordingly, leveraging historical insights to inform future strategies. On the other hand, a prediction is an actual act of indicating that something will happen in the future with or without prior information. For example, speculating annual personnel turnover in the coming 12 months, based on previous data, is a prediction. The second concept is about adopting a mindset of learning from failure. The idea here is that probes that don't work also contain valuable information about what is occurring.

FORECASTING

Forecasts present most likely futures, based on assumed drivers for change. They feature

- a limited range of variation
- are based on a given set of drivers
- rely on historical data in combination with conjectures of future development.

An advantage of forecasting methods and at the same time their drawback is that they tend to treat the future as a form of continuation of the past. As discussed in Chapter 14, taking the past as a guide to the future is a drawback as it hides the inherent uncertainty of the future, especially regarding disruptive changes. In a nutshell, developing a forecast is best for an environment that has minimal uncertainty, whereas scenario planning (discussed later in this chapter) aims to speculate on multiple futures.

In fact, although many organizations pay lip-service to foresight, not many engage it in a proactive way to expand thinking about potential futures. In a survey of over 400 executives from Forbes Global 2,000 firms, Schwarz and Kleine (2025) found that, while the notion of strategic foresight was well accepted within firms, fewer

DOI: 10.1201/9781003566007-18

than half reported using it in explorative or adaptive ways, and many still preferred established strategies over innovation. In short, the value of foresight in dynamic VUCA-related conditions is under utilized.

DEVELOPING FORESIGHT

Silver (2012) quotes psychologist Philip Tetlock, who examined the cognitive thinking styles of experts claiming that experts are not very good at prediction because they are typically overconfident in their own expertise and poor at calculating probabilities. He classified the styles on a spectrum between what he called the hedgehog and the fox.

According to Silver (2012), hedgehogs are characterized as Type A personalities who adhere to overarching "big ideas" or governing principles that they perceive as fundamental truths guiding virtually every interaction within society (e.g., Karl Marx, who emphasized class struggle, or Sigmund Freud, who focused on the role of the unconscious in human behavior).

In contrast, foxes are claimed to be more adaptable and resourceful. They embrace a variety of smaller ideas and adopt multiple approaches when tackling problems (Silver, 2012). This viewpoint posits that foxes tend to navigate nuances, uncertainties, complexities, and differing viewpoints with greater ease, displaying a more tolerant attitude toward dissenting opinions. This flexibility allows them to engage more dynamically with the complexities of the world around them.

In a later book, Tetlock and Gardner (2016) contend that foxes are considerably better at forecasting than hedgehogs. While there is a debate about whether it is reasonable to classify foxes and hedgehog-type personalities in this way (see e.g. Dworkin, 2011), the pragmatic question for practitioners operating in dynamic situations is: how might they be more like foxes in forecasting to develop better plans?

> For super-forecasters, beliefs are hypotheses to be tested, not treasures to be guarded.
> *(Tetlock & Gardner, 2016, p. 127)*

One way to become more like foxes when forecasting is to adopt a fox-like mindset. The following is adapted from Tetlock and Gardner (2016), who suggest foxes can become super forecasters when they practice the mindsets included in Table 15.1.

TABLE 15.1
Mindsets of Foxes

Cautious	Nothing is certain
Humble	Reality is infinitely complex
Non-deterministic	What happens is not meant to be and does not have to happen
Actively open minded	Beliefs are hypotheses to be tested not treasures to be guarded
Intelligent and curious	Intellectually curious, enjoy puzzles and mental challenges
Reflective	Introspective and self-critical
Numerate	Comfortable with numbers and estimates

Source: Tetlock and Gardner (2016).

STRATEGIES FOR PRACTITIONERS: APPLYING FORESIGHT TO ENHANCE FORECASTING

According to Saffo (2007), there are six rules for effective foresight.

Rule 1: Define and Map a Cone of Uncertainty

- **Delineate Possibilities:** Begin by outlining a range of possibilities that extend from a specific moment or event. This involves defining the breadth of the cone, which represents the overall uncertainty involved. In addition to this, it is important to examine the relationships among various elements and prioritize possible outcomes while developing a forecast. Establishing the breadth of the cone is the crucial first step in this process.
- **Identify Wild Cards:** Consider the outliers or "wild cards" – outcomes that may occur but are often uncomfortable to think about. These events should be sufficiently surprising or unconventional to capture attention, but you also need to ensure they remain relatable and don't alienate your audience.
- **Avoid a Too-Narrow Cone:** Be cautious not to draw the cone too narrowly, as this scenario can be more problematic than making it too broad. Ensure you don't concentrate solely on the edges at the expense of the center; failing to consider the central elements can lead to missed certainties. Above all, challenge whether certain wild cards might deserve to be repositioned closer to the center based on emerging information or insights.

Rule 2: Look for the S-curve

- **Recognize the Nature of Change:** Change seldom progresses in a linear fashion. Most significant developments tend to follow an S-curve pattern, characterized by a slow initial phase, followed by a period of rapid acceleration, and ultimately a tapering off or even a decline. This reflects the typical power law associated with transformative changes.
- **Identify Emerging Patterns:** The key to effective foresight is to spot the emergence of an S-curve pattern well before reaching the inflection point. This early identification allows for proactive planning and adaptation.
- **Beware of Short-Term Optimism:** Avoid confusing a clear perspective for a short horizon. There is a common tendency to overestimate immediate outcomes while underestimating long-term developments. Our aspirations often lead us to believe that significant transformations will happen overnight. When reality contradicts this belief, we may become disillusioned, falsely concluding that the anticipated change will never materialize – just before it actually does.
- **Acknowledge the Pace of Change:** Be mindful of the speed at which change can occur. Even when we recognize the onset of a transformation, our instinct may lead us to draw a straight line through the S-curve. While we may eventually reach the same endpoint, this linear perspective can cause us to overlook the initial delay as well as the explosive growth that follows in the middle stage. Remember, timing is crucial.

Rule 3: Embrace the Things that don't Fit

- **Tune into Anomalies:** To effectively identify an emerging S-curve, forecasters should develop a sensitivity to elements that do not conform to established patterns or categories. These oddities, which people might dismiss or reject, can offer valuable insights. Our natural inclination to seek certainty and focus on the present often leads us to overlook these indicators.
- **Recognize the Value of the Unfamiliar:** Anything that is genuinely innovative or groundbreaking is unlikely to fit neatly into existing categories. Embracing this truth is essential for developing foresight.
- **Reframe Failures and Oddities:** Many indicators may initially present themselves as strange curiosities or even as failures. Understanding that these anomalies can signal potential shifts or emerging trends is vital. Our aversion to uncertainty may lead us to shy away from these indicators, but they often hold the key to anticipating significant changes.

Rule 4: Avoid Overreliance on Strong Information

- **Beware of Confirmation Bias:** One of the most significant pitfalls for practitioners is the tendency to place excessive trust in a single piece of compelling information (i.e., a hedgehog view) particularly if it aligns with their preexisting conclusions. This confirmation bias can lead to skewed perspectives and poor decision-making.
- **Value Interconnected Weak Information:** A multitude of interlocking weak indicators tends to be far more reliable than one or two strong data points. Unfortunately, traditional research methodologies often prioritize the collection of robust evidence, leading to a narrow focus. Once researchers invest time in developing a well-crafted hypothesis, they may be inclined to dismiss any contradictory evidence.
- **Understand Resistance to Contradictory Evidence:** This reluctance to accept opposing viewpoints contributes significantly to the nonlinear nature of paradigm shifts, as described by Thomas Kuhn in his seminal work, The Structure of Scientific Revolutions (1970).
- **Embrace Strong Opinions, Weakly Held:** Foresight relies on maintaining strong opinions that are not rigidly held. If you engage in forecasting, do so frequently and remain open to the possibility of being proven wrong. This approach fosters flexibility and adaptability, allowing you to adjust your views based on new data and emerging insights.

Rule 5: Look Back Twice as far as You Look Forward

- **Value Historical Perspective:** Marshall McLuhan (2012) noted that people often navigate their future while fixated on the past, as history tends to feel more reassuring than the present. While this inclination can mislead, when applied correctly, examining past events can be a potent tool. The nuances and patterns of historical events can help make sense of current indicators and effectively project future trends, but it is crucial to look back far enough.

- **Focus on Patterns, Not Just Events:** When analyzing historical data, concentrate on identifying significant shifts or turns rather than merely the straightforward paths. This deeper exploration allows for the recognition of underlying patterns that can inform future scenarios.
- **Adopt a Double-Interval Approach:** When seeking parallels in history, always look back at least twice as far as you plan to look forward. This approach increases the likelihood of uncovering relevant patterns. Keep in mind that while history can offer valuable insights, it rarely repeats itself exactly. If a double interval doesn't provide enough context to inform your intuition, don't hesitate to delve even further back.
- **Learn from Past Conflicts:** A humorous admonition at the Pentagon is that the US military tends to "fight the last war". This was evident during the Iraq conflict when some planners assumed that Iraq II would mirror Iraq I, while simultaneously neglecting lessons from Vietnam because of the perception that it was a war that was lost. Such examples underscore the importance of extracting lessons from a comprehensive historical analysis.

Rule 6: Know when Not to Make a Forecast

- **Understand the Complexity of Change:** As humans, we have a unique relationship with change, simultaneously feeling both fear and fascination. In times of rapid transformation, it is essential to recognize that many aspects remain constant, often outnumbering the new developments that arise.
- **Acknowledge the Dynamic Nature of the Cone of Uncertainty:** The cone of uncertainty is not fixed; it fluctuates as the present transitions into the future. As specific possibilities unfold, others become less likely. During periods of extreme uncertainty, the cone can expand significantly, making it prudent for forecasters to hold off on making predictions altogether. However, it is reassuring to remember that stability will eventually return, allowing for more informed forecasting once again.
- **Embrace Skepticism Toward Change:** Exercise caution with apparent changes and avoid rushing into predictions during turbulent periods. It is wise to refrain from placing too much weight on any single forecast during such times. Instead, recognize that a wave of new indicators will emerge, providing further insights sooner than anticipated.

SCENARIO PLANNING

One way of identifying potential trends is with scenario planning. This enables practitioners to distinguish the developments that are more likely to become trends and separate these from the noise.

The scenario planning process is defined as the means of "creating stories of equally plausible futures and planning as though anyone could move forward" (Tucker, 1999, p. 70, cited in Burt & Nair, 2020). Scenarios are regarded as a hypothetical sequence of events that may lead to a possible future. The key idea here is that scenarios are not predictions, as the following quote from the

Commonwealth Scientific and Industrial Research Organization's (CSIRO's) Futures report (2016) articulates.

> The value of developing scenarios is not in deciding which is most likely, but rather in contrasting what each scenario can mean.
>
> (*CSIRO, 2016, p. 9*)

Developing Scenarios

Scenarios are alternate futures in which today's decisions may play out. They are stories with beginnings, middles, and endings. Good scenarios have twists and turns that show how the environment might change over time (Sandberg, 2018). According to Sandberg, it is worth considering the following principles in relation to scenarios:

- Scenarios are hypotheses of the future, not predictions.
- Scenarios encourage divergent thinking by design.
- Scenarios set the stage to plan for multiple futures.
- Scenarios are rich, data-driven stories about tomorrow that can help you make better decisions today.

According to Sandberg (2018), to work properly, four to five scenarios need to be developed. More than five scenarios tend to get confused with one another. Three scenarios run the danger that people will try to pick the most moderate or most apparently plausible and forget about the other two. Four is a good number – neither too many nor too few.

Alternate scenarios can serve as relatively low-cost insurance policies. You are less likely to be blindsided if you've taken the trouble to imagine some unwelcome surprises. However, there is still a problem. Not all scenarios have the same quality to advance critical and creative thinking about potential futures.

Enhancing the Quality of Scenario Planning

As Silver (2012) points out, we focus on stories about the world as we would like it to be. Moreover, the approximations we make are often much cruder than we realize.

Curnin et al. (2025) contend that fostering critical thinking skills is important, citing literature demonstrating that much scenario planning lacks creativity. To encourage creativity and critical thinking in scenario planning, they worked with teams to first ensure psychological safety and swift trust because it takes courage to voice unusual ideas (see Chapter 9). In building scenarios, Curnin et al. encouraged teams to use a variety of techniques including SCAMPER (see Table 15.2) to develop divergent thinking.

In this approach, teams brainstorm different potentialities and, once exhausted, extend their thinking by applying different attributes in SCAMPER to reframe potential futures. For example, if teams had identified two consequences (for example, when using ripple mapping approach described in Chapter 12) of the potential scenario, how might these two consequences combine to create an opportunity?

In their research Curnin et al. (2025) demonstrated significant enhancements in speed and quality of the scenarios developed when they employed SCAMPER

TABLE 15.2
The SCAMPER Technique

Theme	Guide
Substitute (e.g., components, materials, people)	Think about substituting part of the product or process for something else. Typical questions: What else instead? Who else instead? What other materials, ingredients, processes, power, sounds, approaches, or forces might I substitute? Which other place?
Combine (e.g., mix, combine with other assemblies or services, integrate)	Think about combining two or more parts of the product or process to make something new or to enhance synergy. Typical questions: What mix, assortment, alloy, or ensemble might I blend? What ideas, purposes, units, or appeals might I combine?
Adapt (e.g., alter, change function, use part of another element)	Think about which parts of the product or process could be adapted or how you might change the nature of the product or process. Typical questions: Does the past offer a parallel? What else is like this? What other idea does this suggest? What might I adapt for use as a solution? What might I copy? Who might I emulate?
Magnify/Modify (e.g., increase or reduce in scale, change shape, modify attributes)	Think about changing part or all of the product or process, or distorting it in an unusual way. Typical questions: What other meaning, color, motion, sound, smell, form, or shape might I adopt? What might I add?
Put (e.g., Put to other uses)	Think of how you might put the product or process to another use or how you might reuse something from somewhere else. Typical questions: What new ways are there to use this? Might this be used in other places? Which other people might I reach? To what other uses might this be put if it is modified?
Eliminate (e.g., remove elements, simplify, reduce to core functionality)	Think of what might happen if you eliminated parts of the product or process and consider what you might do in that situation. Typical questions: What might I understate? What might I eliminate? What might I streamline? What might I make smaller, lower, shorter, or lighter?
Rearrange/Reverse (e.g., turn inside out or upside down)	Think of what you might do if parts of the product or process worked in reverse or were sequenced differently. Typical questions: What might be rearranged? What other pattern, layout, or sequence might I adopt? Can components be interchanged? Should I change pace or schedule? Can positives and negatives be swapped? Could roles be reversed?

Source: Serrat (2017).

as one of the tools to develop fast and frugal scenario planning in crisis management teams.

In our own work, we found that participants could extend their creative thinking beyond the obvious in scenario planning if we engaged them in drawing their potential scenarios (see Figure 15.1a–c). This came with some rules requiring each member to hold at least 2 marker pens and for all to work simultaneously across their respective scenario imagery displayed on three A2 sticky sheets on a wall. The participants were then required to explain their storylines to others, resulting in much richer narratives than if they simply wrote them down.

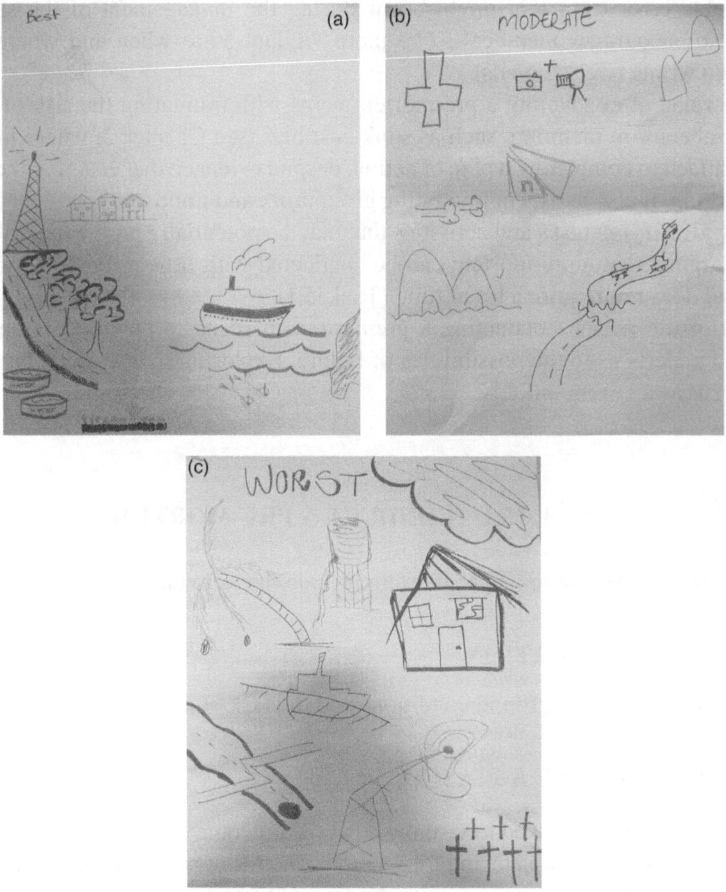

FIGURE 15.1 (a–c) Participant drawings of best, moderate, and worst-case scenario outcomes to illustrate three possible scenarios to a case study of a disaster event.

Using Scenarios in Planning for Contingencies

One strategy that elaborates on scenario planning is called the pre-mortem. A post-mortem is an investigation of someone who has died to understand the cause. In a pre-mortem, decision-makers consider ahead of time what could possibly go wrong to identify what contingencies can be put into place to improve the robustness of a plan. This makes a pre-mortem a generic strategy for facilitating error detection. Claire Johnson (2017) has modified the pre-mortem approach to facilitate error detection in a wide range of VUCA-related situations. The core of the tool is that it allows participants to engage in a process of imagining the future to imagine things that might go wrong (i.e., also known as prospective hindsight).

According to Johnson (2017), undertaking a pre-mortem in VUCA contexts supports practitioners to bring together the collective knowledge in the group as each person has a different set of experiences, and a different mental model to bring to the task of executing a plan. Doing a pre-mortem also helps sensitize members to possible error traps that might occur during the management of the situation. This also encourages members to be more vigilant as to when and where things might go wrong (weak signals).

The value of conducting a pre-mortem helps with mitigating the risk of certain biases inherent in planning, such as sunk-cost bias (see Chapter 2) where members overly attach to completing a plan of action, despite evidence that enacting the plan is not working. It also assists in mitigating groupthink and improves contingency planning by identifying tasks and activities that may be potentially problematic. Johnson (2017) argued that a pre-mortem can be conducted with little advance notice/planning and does not require a lot of time (Tool 15.1).

Part of the reason conducting a premortem in scenario planning is helpful is because it alerts us to the possibilities of failure. Exploring indicators of failure can also occur as an event unfolds.

TOOL 15.1 CONDUCT A PRE-MORTEM

The steps to conducting a pre-mortem include the following:

STEP 1 – PREPARATION

Summarize the background conditions and the task objectives of the team.

STEP 2 – IMAGINE A FIASCO

- Members are asked to visualize that the outcome of enacting the plan has been a complete and total fiasco. The result might be embarrassing and devastating – sometimes called "front page news".
- Members consider they have a crystal ball looking into the future that can show why the plan went spectacularly badly. However, the crystal ball is a cheap one, and so members cannot see why things went catastrophically wrong.
- The team is encouraged to consider as many answers as possible to the why things went catastrophically wrong and not just to find one "right" answer. That is, the focus is not on attempting to "guess the outcome" or to criticize those who were in the situation but on the many possible ways that could have gone wrong.

STEP 3 – GENERATE REASONS FOR FAILURE

- The team are then encouraged to spend three minutes individually writing down as many reasons as to why the failure occurred. Members are reminded that attention is not so much to what the failure was but rather on why the failure happened.

- Once individuals have thought about it independently then the ideas are discussed with other members. Members are encouraged to be realistic and at the same time, to be creative. For example, there could be many different timeframes involved as well as events that are both inside and outside of the team's control.

STEP 4 – CONSOLIDATE THE LISTS

- Members discuss and record all reasons for failure in the group and brainstorm possible convergence and divergence of ideas.
- The team is also asked to consider any particular issues that are more critical than others.

STEP 5 – REVISIT THE PLAN

- The plan to address elements in the VUCA-related situation is then revisited in light of these possible failures and adjusted where required to make it more robust to failure.

The whole point of pre-mortems is not merely to identify possible worst-case outcomes to strengthen and improve a plan, but in doing so to create a climate that facilitates error trapping.

Failure Management

In probing to make sense of a situation failure often occurs. Indeed, the word "fail" as an acronym has been popularly quoted as "first attempt in learning" by APJ Abdul Kalam.

According to Lee (2018, p. 166), failure is "a state in which reality is inferior to one's goals or expectations". Managing failure can be either a retrospective or a prospective activity in the face of failure. Failure management means that failure can and should be analyzed deliberately to gain new knowledge pertaining to the causes of failure and pursue new opportunities beyond the current approach. This means that the way we approach failure and how we learn from it is important.

Simply put, the preparedness of failure management is the capability to learn and make constructive decisions in the face of failure. Do we anticipate failure in advance? Is there anticipation of what failure will look like? Ask these questions

- How is failure possible?
- What are the signals failure is occurring?
- How can these be monitored before the failure occurs?
- Have the benefits of failure been anticipated in advance?

However, failure can often bring on the "blame game" so a nuanced understanding of failure is needed.

Experimentalism and Iterative Learning

The very nature of complexity in crisis for decision-makers under pressure is the need to create situations where "safe-to-fail" experiments are enabled to allow innovative ideas to germinate and emerge in safe situations, and patterns to surface. This requires leadership and guidance.

The purpose here is to create "safe to fail" probes that can be used to better understand the patterns occurring in complex and dynamic environments. According to the Cynefin site,[1] safe-fail probes are small-scale experiments that approach issues from different angles, in small and safe-to-fail ways, the intent of which is to approach issues in small, contained ways to allow emergent possibilities to become more visible. Boundaries are clearly important. These include

- Being comfortable with "safe uncertainty" – that is with providing just enough structure and control to mitigate the biggest of risks while leaving sufficient fluidity, spontaneity, and freedom to welcome new possibilities.
- Making experiments safe to fail – by ensuring you can afford the experiment to fail.
- Designing experiments that can be measured or observed in their outcomes – knowing what the indicators are of success or failure.
- If you can do multiple experiments on the same system at the same time – some will work, some won't. Shut down the unsuccessful ones and create variations on the ones that work.
- Be prepared for surprise.
- Don't explain unexpected effects away as noise or the exception to the rule but learn from them.

The emphasis, then, is not on ensuring success or avoiding failure, but in allowing ideas that are not useful to fail in small, contained, and tolerable ways and discover more about the system in the process. Safe-fail probes enable practitioners to make progress on adaptive challenges. Developing an experimental mindset can be facilitated by developing a portfolio of probes. These can include:

- Encouraging thinking outside of traditional boundaries. Safe-fail probes require and enhance the ability for exploration.
- Working through some interesting examples of oblique solutions that illustrate the importance of not thinking in direct, causal relationships, preferably from a different domain than the one you are operating in.
- Promoting cross-silo and cross-rank collaboration and sensitization.

In facilitating "safe-to-fail" probes, the Cynefin site suggests not providing the team with direct examples relating directly to their own context, as this has the effect of patterning how team members start thinking about their problem.

Oblique examples from different industries are best for conveying the principle behind this method.

MAIN POINTS

- Developing foresight requires developing a mindset that is actively open-minded and curious, reflective, non-deterministic, and humble.
- When applying foresight, it is useful to recognize that change often follows an S-curve pattern: that is, there is a slow initial phase, rapid acceleration, and tapering off or decline. Early identification of emerging S-curve patterns allows for proactive planning and adaptation.
- In applying forecasting, it is helpful to avoid short-term optimism as there is a tendency to overestimate immediate outcomes and underestimate long-term developments.
- Tune into anomalies that do not fit established patterns, as they can provide valuable insights for forecasting.
- Recognize that innovative ideas may not fit neatly into existing categories and embrace these unfamiliar elements for accurate forecasting.
- Scenarios are not predictions – they are storylines of plausible futures, any of which might move forward.
- A scenario planning process, known as a pre-mortem, can be used to evaluate the robustness of a plan and to critically assess its potential failure points. Doing so enables decision-makers to engage in improving the robustness of planned actions.

NOTE

1. https://cynefin.io/wiki/Safe_to_fail_probes

REFERENCES

Burt, G., & Nair, A. K. (2020). Rigidities of imagination in scenario planning: Strategic foresight through 'Unlearning'. *Technological Forecasting and Social Change, 153*, 119927.

CSIRO (2016, May). *Australia 2030 – Navigating our uncertain future*. CSIRO. Clayton. CSIRO: EP161882. https://doi.org/10.4225/08/58557dbc66880

Curnin, S., Brooks, B., Brooks, M. O., & Akhtar, M. S. A. (2025). Learning for uncertain futures: Training teams to rapidly develop scenarios during organisational crises. *Futures, 173*, 103651.

Dworkin, R. (2011). *Justice for hedgehogs*. Harvard University Press.

Johnson, C. (2017). Expert decision making and the use of worst-case scenario thinking. In C. Owen (Ed.), *Human factors challenges in emergency management* (pp. 35–56). CRC Press.

Kuhn, T. S. (1970) *The structure of scientific revolutions* (3rd ed.). University of Chicago Press.

Lee, J. (2018). Making hindsight foresight: Strategies and preparedness of failure management. *Organizational Dynamics, 47*(3), 165–173.

McLuhan, M. (2012). The medium is the message. In *Media and cultural studies: keyworks* (pp. 100–107).

Saffo, P. (2007). Six rules for effective forecasting. *HBR, July-August.* Retrieved June 2, 2025, from https://hbr.org/2007/07/six-rules-for-effective-forecasting

Sandberg, K. (2016). *Scenario planning.* medium.com/strategy-dynamics/scenario-planning-part-1-ffcb48f26a1c

Sandberg, K. (2018, Jan 6). Scenario planning: Part 1. *Medium.* Retrieved June 2, 2025, from https://medium.com/strategy-dynamics/scenario-planning-part-1-ffcb48f26a1c

Schwarz, J. D., & Kleine, K. (2025) The state of Corporate Foresight: Global study of the Application of Corporate Foresight, report Bayerisches Foresight-Intitiut. Available https://www.thi.de/fileadmin/daten/forschung/Foresight/SHOTS/TheStateofCorporateForesight.pdf

Serrat, O. (2017). The SCAMPER technique. In *Knowledge solutions.* Springer. https://doi.org/10.1007/978-981-10-0983-9_33

Silver, N. (2012). *The signal and the noise: Why so many predictions fail-but some don't.* Penguin.

Tetlock, P. E., & Gardner, D. (2016). *Superforecasting: The art and science of prediction.* Random House.

Tucker, K. (1999). Scenario planning: Visualizing a broader world of possibilities can help associations anticipate and prepare for change. *Association Management, 51*(4), 70–77.

Whitehead, A. N. (1947). *Adventures of ideas.* CUP Archive.

16 Psycho-social Behavior in Crises

INTRODUCTION

As a psychotherapist of grief and recovery, Gordon (2004, 2006) has worked with many communities affected by VUCA-related disaster events. For Gordon, the defining element of such situations is that they happen to members of a community. Through his observations, he has identified several common themes that emerge as communities live through and then recover from these crisis situations. Better understanding of how the public is likely to behave during and following trauma assists practitioners to engage with community members to move forward.

Gordon proposes that we don't fully appreciate how much of our lives are underpinned and supported by routines of community life. This is because these social processes are taken for granted. A high-consequence VUCA situation such as a disaster is, among other things, a disruption to these routines. In so doing, the disaster stimulates a range of stresses on social life. Gordon (2006) suggests that it is important for decision-makers and those in government positions to always keep the psycho-social aspects in the foreground and not get overly focused by the physical destruction.

For people experiencing crises, community members are simultaneously experiencing a series of intense emotions (see also Chapters 3 and 4). Experiencing a disaster can lead to a range of psychological and cognitive effects, including high arousal, shock, narrowed focus, disbelief, fatigue, exhaustion, apathy, and confusion regarding the events that have occurred or are occurring. According to the Australian Red Cross (2022), the impact of a disaster can significantly affect an individual's abilities in several key areas:

Concentration:

- *Amount of New Information:* The capacity to absorb and comprehend new information can be severely limited.
- *Complexity of Detail:* Individuals may struggle to process complex details or nuances.
- *Duration of Focus*: The length of time a person can maintain focus is often decreased.

Memory:

- *Recall of Information:* The ability to remember spoken, written, or visual information can be impaired.
- *Simplicity vs. Complexity*: Recalling simple or complex knowledge becomes more difficult.
- *Timeframe of Memories*: Individuals may have trouble recalling both recent and historical memories.

DOI: 10.1201/9781003566007-19

Decision-Making:

- *Evaluating Risks:* Weighing possibilities and assessing risks can become challenging.
- *Handling Complexity*: Individuals may find it difficult to manage complex ideas and potential outcomes.
- *Action Planning:* Prioritizing and planning actions may be hindered due to cognitive overload and confusion.

These effects highlight the importance of providing support and clarity to community members experiencing crisis situations, as their cognitive capabilities may be significantly compromised.

EXPERIENCING PSYCHO-SOCIAL DISRUPTION

Gordon (2006) has observed that members of communities move through a range of processes, and what decision-makers and officials do within these phases impacts the degree to which that community will be psycho-socially impacted. Table 16.1 shows the common phases Gordon has observed, and some of the strategies practitioners can implement.

According to Gordon (2006), human behavior during a crisis goes through several stages, and he calls these the social functions: (i) before the disaster, (ii) survival mode during the crisis, (iii) fusion, (iv) cleavage, and then (v) the re-creation of a new individual and social identity.

i. *Pre-disaster*
The community is a complex, structured environment, consisting of routines, social roles, and relationships that allow people to have a unique combination of similarities and differences. Each person has their own status, position, and attachments to people and places.

ii. *Survival mode*
When a crisis occurs, it puts everyone out of their normal routines and into survival mode – that is, a momentary shift where people drop out of the social fabric to face the threat. To face the threat, members need to abandon all the routine connections that preserve their normal life. This is a process Gordon (2006) calls "debonding". This lasts for as long as members directly face the threat. In other words, what individuals are experiencing can be summed up as – *"it is me and the threat, and I will help others where I can"*.

iii. *Fusion*
As soon as survival is not in question, people explode into action to start dealing with the consequences. However, this is not the social fabric of their normal life, but rather a highly altered social process where their roles and priorities are all improvised around immediate demands. There is a loss of social roles and relationships, and an improvisation of roles defined by the immediate physical requirements.

iv. *Cleavage*
However, the intensity of fusion is not sustainable as it does not allow for the needs of normal life. There is a gradual realization that we are not necessarily "all in this together". A tension emerges: how do community members get from tight, high-energy fusion back to the complexity of routinized social fabric? The differences in the consequences of the crisis become manifest (e.g., the person who is struggling with insurance claims and the person receiving a benefit who has no insurance). Cleavage brings with it negative emotions, hostility, envy, conflict, and competition.

v. *Re-creation of a new identity*
According to Gordon (2006), recovery is not a community going back to what it was previously, but what comes when the community re-establishes a community identity again. Recovery is complete when community members are living the life they value.

Working with Communities

Outsiders arriving in a disaster or crisis-affected community often see community members in survival mode, or a "de-bonded" state. Gordon (2006) observes that people may appear disoriented and often in a bewildered or very excited state. He observes that this does not mean they are dysfunctional. It means that they have dropped out of the normal social fabric. Gordon proposes that the length of time people exist in that highly aroused, de-bonded state is a strong predictor of how severe their psycho-social trauma will be later. How the practitioners involved try to re-establish normality and routines is very important. Gordon suggests this means minimizing an authoritative approach. He suggests decision-makers and authorities talk with community members quietly, consulting with them, even if that takes a little longer.

The transition from a de-bonded state to a state of fusion releases an enormous amount of energy. People often have a sense of euphoria, laughing, joking, and hugging each other. According to Gordon (2006), this is a temporary state and is not reflective of the long-term impacts of the tragedy that has happened. This is, however, the period during which people reattach to one another. Gordon suggests that practitioners need to establish themselves during the fusion stage as soon as possible.

People are in a highly aroused, psychological state, having come out of the de-bonding stage, and are seeking to attach to whatever is important to them, commonly the people who are around them. If representatives from various response and recovery agencies are not there, then the risk is that the community won't attach to them. In which case, when personnel from the various organizations do arrive, there will be resentment – *"Where were you? Why has it taken so long? Do you really care? We've been looking after each other, we don't need outsiders here"* – even though this is unrealistic. Gordon (2006) suggests that practitioners aren't required to do much at this stage, other than convey to the community what is happening and what is being organized. This is the point at which communication and consultation with the community can and should begin.

The cleavage stage is a more difficult stage for practitioners to navigate. Practitioners face being the recipients of the negative emotions that those engaged in the cleavage plane stage might be feeling. A strong sense of one's own emotional intelligence and emotional social awareness will be important here (see Chapter 5). Decision-makers need to have a strategy in place to facilitate moving through these emotions. It is also important that the practitioner recognizes that while the community might look ostensibly like they are carrying on functioning, they are operating, however, with an altered quality of life, which is ultimately unsustainable.

Facilitating normal social routines (e.g., community barbecues and other leisure activities) will assist a community moving through this phase. Gordon also notes that community and organizational memberships are likely to have significant turnover at this time. Promoting opportunities for the community to come together and to tell their stories if they wish helps community members to acknowledge the community's losses but also their resilience, with the hope that the community comes out of

TOOL 16.1 ASSESSING POTENTIAL FOR SOCIAL DE-BONDING

Social de-bonding is a process of social disconnection. "De-bonding" accompanies a profound disruption to the pre-existing continuity of physical, emotional, and social life (Gordon, 2004). In considering the impacts of disaster on the social function of a community, it is useful to reflect on the following indicators.

TABLE 16.1
Indicators of the Impact of a Crisis on Social Functioning

Intensity – How profound and violating is/was the threat? To what degree is/was it related to the subjective sense of imminent death, injury or helplessness?					
High likelihood of imminent death/injury		↔			Low likelihood of imminent death/injury
Pervasiveness – How immersed were people in the situation, did the whole world seem engulfed, or was the threat specific and more contained?					
Whole world engulfed		↔			Specific and contained
Duration – How long did the threat persist? How long before safety and security were re-established?					
Long duration		↔			Short duration

Source: Gordon (2004).

the event with the social benefits of having learned together and connected in ways they weren't before.

Tool 16.1 has been developed following the work of Gordon (2006) as a subjective assessment of the potential for social debonding, as this is critical to understand and guide the approaches practitioners working within disrupted communities can take.

NAVIGATING IN TIMES OF PSYCHO-SOCIAL CRISIS

It is important to recognize the extent to which people and their communities were functioning prior to the crisis and before the need for disaster support arrangements. Practitioners can be mindful that the community previously operated as a self-governing network of people and is not represented by what the practitioner initially sees and confronts. How the community functioned prior to the disaster, and the existing networks should be known. However, for those appointed from outside the local area, this will require these practitioners to reach out and learn about the physical, psychological, and social resources available within the community. In a nutshell, a practitioner must leap into the unknown – to try and find out what was here before.

Finding Information Conduits and Trusted Voices

One way to try and make sense of what is happening with the community is to look for information conduits and trusted voices (Australian Red Cross, 2022). These are the key individuals within the community who can play a vital role in the analysis, distribution, and dissemination of information. These can be formal or informal community leaders and members who possess a deep understanding of the community's needs and are well-connected with its people. Engaging these individuals can help practitioners achieve several important objectives:

- **Bypass Informational Noise:** Engaging these trusted sources can help filter out irrelevant information and reduce confusion within the community.
- **Target Communications Effectively:** These individuals can provide insight on how to tailor messages more accurately to address the specific needs and concerns of the community.
- **Ensure Accurate Information Reach:** Leaders can have increased confidence that the information shared is reaching the intended audience effectively.
- **Improve Feedback Mechanisms:** Working with knowledgeable community members can lead to more valuable feedback about the effectiveness of communications.
- **Understand Information Needs:** They can help identify what information is essential for the community and when it is needed.

When reaching out to these key individuals, it is important to incorporate strong feedback loops into communication efforts. Many people conceive communication only as the act of sending information, but this is only part of the process. A successful information campaign requires the establishment of mechanisms to gather feedback, suggestions, and questions from the community audience. The advantages of feedback loops include the ability to:

- **Gain Insights:** Understanding specific needs and issues prevalent within the community helps tailor responses accordingly.
- **Evaluate Communications:** Continuous feedback allows for ongoing review and assessment of messaging and service effectiveness.
- **Tailor Information:** Adapt communications to meet the community's evolving needs based on actual feedback.
- **Assess Relevance:** Determine whether communication efforts are valuable or if adjustments are necessary to enhance relevance in the future.

While building strong feedback loops into communication strategies can be challenging and time-consuming, engaging with the audience to receive their feedback will ultimately enhance service delivery in the long run. Practitioners should actively consider how to facilitate the reception and application of feedback throughout the crisis recovery process.

STRATEGIES FOR PRACTITIONERS

Gordon (2004, 2006) outlines several strategies that practitioners can implement to mitigate the negative impacts of de-bonding and social disruption in the aftermath of a disaster. These initiatives focus on fostering communication, community engagement, and social cohesion. Below is a summary of the strategies he proposes practitioners can facilitate to overcome the debilitating psycho-social effects of crises:

Re-bonding

- *Establish Communication Channels*: Initiate communication with the community and maintain connections among residents and key stakeholders as promptly as possible. This helps discourage isolation and encourages ongoing interaction.

Community formation

- *Define Affected Groups:* Quickly identify who has been affected by the disaster and ensure that members are aware of each other collectively, fostering a sense of community and shared experiences regarding their challenges and needs.

Facilitate social bonds

- *Create Unified Communication Systems:* Implement platforms, such as a community newsletter, to provide updates on the recovery process and set realistic expectations, helping to unify the affected community.

Normalize

- *Normalize Disaster Effects:* As early as possible, normalize the emotional and psychological effects of the disaster. Sharing personal anecdotes can promote open communication about experiences and feelings within the community.

Form disaster-related social representations

- *Encourage Dialogue about Experiences:* Facilitate discussions within the community that help articulate shared experiences and reactions to the disaster, preparing individuals for common challenges while reinforcing mutual understanding.

Form a common reality

- *Provide Accurate Information*: Be proactive in sharing information regarding the disaster, including its causes and implications, to reduce uncertainty and clear up any misconceptions that may arise.

Form a frame of reference

- *Establish Informational Foundations*: Create a robust body of information that can guide informed evaluations regarding the disaster and responses, helping community members make sense of their situation.

Preserve differences and complexity

- *Encourage Diversity of Expression*: Actively combat tendencies toward homogenization by fostering an environment where individual differences and diverse perspectives are expressed and respected.

Preserve boundaries and identities

- *Contextual Communication:* Ensure that communication occurs within relevant contexts that preserve personal privacy while addressing the matter at hand.

Facilitate reference groups

- *Support Group Formation*: Encourage the formation of both formal and informal groups among individuals facing similar issues, integrating them into the recovery framework through facilitation and resource allocation.

Facilitate social representations of post-disaster life

- *Organize Community Events:* Promote community-driven events that reflect on the disaster and its repercussions, using rituals, symbols, and artistic expressions to foster recovery and remembrance.

Integrate services

- *Align Services with Social Contexts:* Ensure that the introduction of services and support measures aligns with and reinforces social representations of the disaster, taking into account the community's social fabric and dynamics.

Implementing these strategies can help rebuild social connections, foster resilience, and promote a sense of community in the aftermath of a crisis.

MAIN POINTS

- The psycho-social behavior of communities during and after disasters is a complex, dynamic process.
- The stages of Survival Mode, Fusion, Cleavage, and the Re-Creation of a New Identity, highlight the profound impact of the crises on the social fabric of communities.

- Effective navigation requires more than just physical reconstruction; it demands a nuanced understanding of these psycho-social dynamics.
- By adopting strategies that foster communication, community engagement, and social cohesion, practitioners and authorities can mitigate the negative effects of social debonding and facilitate the re-establishment of a community's identity.
- Acknowledging and addressing the psycho-social dimensions of crisis is essential for promoting resilient, thriving communities that can rebuild and flourish in the aftermath of crises.

REFERENCES

Australian Red Cross (ARC) (2022). *Recovery basics: Working in recovery*. Retrieved June 2, 2025, from https://www.redcross.org.au/globalassets/cms/emergency-services/recovery-basics/recovery-basics—working-in-recovery-a5-d5-web-f.pdf

Gordon, R. (2004). The social system as site of disaster impact and resource for recovery. *The Australian Journal of Emergency Management*, *16*(22), 16–22.

Gordon, R. (2006). Acute responses to emergencies: Findings and observations of 20 years in the field. *The Australian Journal of Emergency Management*, *21*(1), 17–21.

17 Navigating the Chaordic Path

INTRODUCTION

When operating in VUCA situations, decision-makers are required to engage in many different roles with a variety of different stakeholders, all serving different functions. As the Institute of Medicine (2015) highlighted, the following stakeholder roles are engaged in support of community well-being and resilience in communities that are impacted by VUCA-related disaster events (see Figure 17.1).

As Figure 17.1 highlights, decision-makers operating in VUCA-related situations are engaged with stakeholders in all four quadrants. However, the relationships they develop will be different and based on different functional role intentions. Most of the activity involved in an operational response will engage organizations and stakeholders in the bottom two quadrants. Practitioners are a conduit between identified government agencies and the affected community and are also responsible for informing political leaders about community needs.

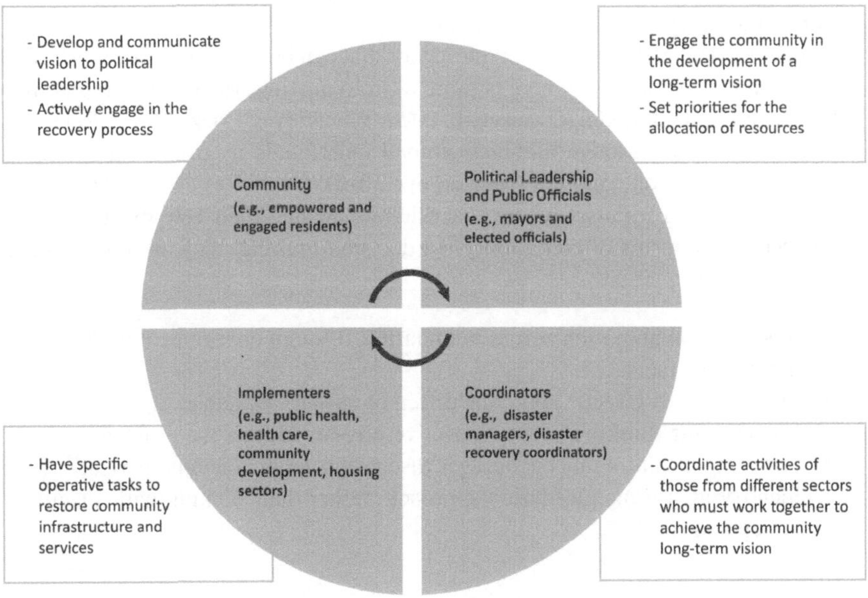

FIGURE 17.1 Key stakeholder roles involved in the disaster recovery process. (From Institute of Medicine (2015). *Communities after disaster: Strategies, opportunities, and planning for recovery*. National Academy of Sciences. https://nap.nationalacademies.org/read/18996/chapter/1#ii)

This means that part of the pressure practitioners working in VUCA-related situations face is that they often need to navigate two worlds:

1. The world of order and bureaucracy, which includes developing plans, applying rules and policies, and achieving outcomes.
2. Living in a chaotic space where there is abundant turbulence, contradiction, and ambiguity.

That is, practitioners need to create environments where there is both stability and order for responding organizations involved to make progress and at the same time sit alongside disaster-impacted community members to help them process their grief, which is often described as living with chaos (see Mitchell, 2019 and Chapter 16). As Petris (2020) contends, empowering individuals and communities to take charge of their recovery journey significantly increases the likelihood of achieving positive outcomes. However, these initiatives can be difficult to get right. There is a balance needed between opportunity and overwhelming affected individuals and communities. Organizations and authorities keen to support can easily undermine their own efforts by operating in ways that disempower.

This requires practitioners to work with communities and their representatives in different ways than they would be working within their own crisis management teams and other stakeholders in managing a response. Practitioners need to mediate between the application of ordered procedures and processes while at the same time facilitating emergent processes that sometimes feel turbulent, contradictory, and messy.

Unlike the leadership style of heroic savior that can become an overused strength (see Chapter 7, the approach here is more akin to hosting. Host leadership is more collaborative and relational (Owen et al., 2024; Wheatley & Frieze, 2011). Relational leadership is the interactive influence among individuals in groups for which the objective is to lead one another (McKergow, 2015). It requires curiosity and a collaborative learning mindset where practitioners change their role by recognizing and accepting in terms of community-led decision-making, they are not in charge and instead are:

- Ensuring collaboration across boundaries, through engagement with community members.
- Stimulating discussion about the different options available.
- Encouraging thinking about innovative ways to address the problem.
- Fostering a culture that encourages discussion about contentious issues.
- Identifying a principles-based approach, rather than sticking with specific policy and rules.

THE ART OF HOSTING

According to Corrigan (2020), the art of hosting requires skills such as those highlighted in Figure 17.2. These include being truly present, engaging skillfully in conversations, being a good host of conversations and engaging with others in cocreation.

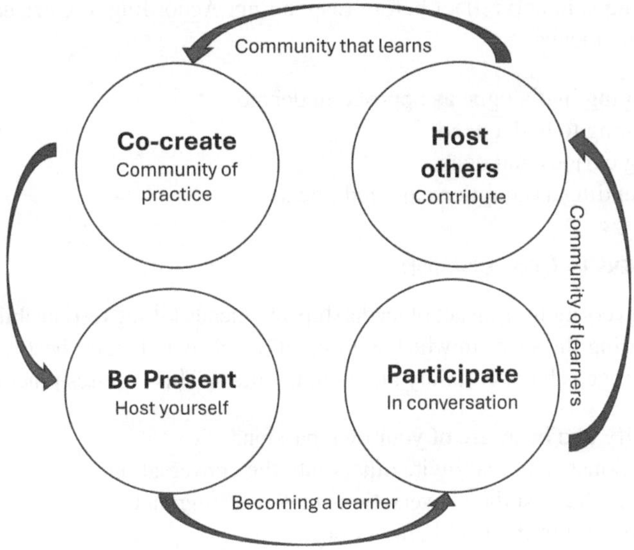

FIGURE 17.2 The four-fold practice of hosting. (From Corrigan, C. (2020). *Reflections on the Four Fold Practice of the Art of Hosting.*)

Be Present

Being present means showing up, undistracted, prepared, clear about the need and what your personal contribution can be. It encompasses the personal practice of curiosity. Presence means making space to devote time dedicated to working with others. If you are distracted, or otherwise located in many different places, you cannot be present. Being present also means being aware of yourself and others (see Chapter 5); what impacts you, and how you impact others. According to Corrigan (2020) in practice, this means finding whatever regular activity helps you to:

- Be present in the moment.
- Create sufficient structure and discipline to focus on what is needed.
- Stay grounded when conflict and chaos arise.
- Be a little more conscious every day.

Participate (Allow Yourself to Be Hosted)

The art of conversation is not just about dialogue. It involves a deep engagement where active listening and genuine curiosity take center stage. When we truly listen to one another, we create an environment that fosters understanding, empathy, and connection. Judgment cannot live in this space. If we are judging, we cannot be curious. Slowing down is essential in this process; it allows us to fully process what is being said and reflect before responding. According to Corrigan (2020) in practice, this means:

- Engaging in dialogue as opposed to debate.
- Listening to understand.
- Being curious and open.
- Suspending assumptions and judgments

Host Others in Conversation

Hosting conversations is an act of leadership and means taking responsibility for creating and holding the space in which a group of people can do their best work together. It takes courage, a bit of certainty and faith in others. This requires practitioners to:

- Identify and be aware of your own passions.
- Consciously call and invite others into the conversation.
- Actively harvest the conversations that are being had.
- Be mindful of the well-being of others.

Co-create

The best conversations happen when we listen for the common ground and what central ideas are evolving from the groups' conversations. In a truly co-creative process, it becomes irrelevant who said or contributed what – the gift is in the synergy and inspiration. Each person's knowledge contributes to the whole, which becomes much bigger than the sum of the parts. This is how results become sustainable over time – when networks of relationships arise from a good conversation. This requires integrating the other three practices to:

- Be open to learn as a team – listening to understand, being curious, suspending assumptions and judgments, and engaging in dialogue as opposed to debate.
- Focus and act on the common ground.
- Consciously seek to build relationships and collective intelligence.

Many of the skills required echo the attributes discussed in earlier chapters; however, there is a nuance here in developing an outward and emergent focus and one where the practitioner is not the one in charge of the decisions being made; but rather handing over the power of decision-making to the community or group. This can be challenging for disaster response practitioners who are used to making decisions. In part, this challenge comes from the need still to be accountable for standards and policies and at the same time to enable a space where community members can take control of the decisions that impact on their lives to reach what Gordon discussed (see Chapter 16) in terms of their re-creation of a new identity.

The emphasis here is on having an outward focus, to work with those who have experienced the VUCA-related situation and to facilitate their sense-making and ownership of the decisions that affect their lives and their communities. Practitioners need to have what Houmann (2022) calls chaordic confidence to persist in support of and to *struggle with* those taking ownership and making decisions themselves. Houmann (2022) says that having chaordic confidence means allowing facilitators to feel comfortable living within order and chaos long enough to support a generative emergence, to live with groups as they struggle with uncertainty, and to resist the temptation to rush for certainty or grab for control.

Tool 17.1 provides some guidance for decision-makers operating in VUCA-related situations to consider when facilitating community-led decision-making.

TOOL 17.1 CONSIDERATIONS WHEN FACILITATING COMMUNITY-LED DECISION-MAKING

When supporting community-led and owned problem-solving and decision-making, practitioners need to:

Avoid:

- Imposing unrealistic time frames.
- Assuming individuals and communities will be available, and willing, to engage.
- Recruiting participants in a way that does not account for potentially marginalized or under-resourced groups.
- Engaging with communities in a "tokenistic" way that does not allow for real influence in decision-making.
- Hiding agendas from community members.
- Asking for feedback with no intention of acting on it.
- Promising the community more than can be delivered.

Consider:

- Giving communities time and space to consider how they would like their work to be carried out.
- Attempting to understand community values and discussions that are already happening (e.g., through informal meetings).
- Identifying and working with local community leaders.
- Encouraging the inclusion of potentially marginalized, or under resourced, groups.
- Being transparent with communities about any constraints or "non-negotiables" (e.g., budgets, legislations).
- Consensus might not always be possible in locally led decision-making.
- Different communities may want to influence decision-making in different ways.
- Monitoring progress and continuously reassessing how communities are being engaged in their own problem-solving and decision-making.
- Finding ways to foster long-term, community-led recovery that will outlast the groups formed at that time.

When communities are empowered, the final decisions are in the hands of community members. It is also worth noting, membership of decision-making groups will vary depending on the event and the associated impacts.

MAIN POINTS

1. Decision-makers engaged in post-VUCA-related recovery situations must engage with various stakeholders across four main spheres: Community, Political Leadership, Implementers, and Coordinators, each serving different functions in disaster response and recovery.
2. In so doing, practitioners face the challenge of navigating two worlds simultaneously: the ordered world of bureaucracy and plans, and the chaotic space of community grief and turbulence following disasters.
3. Effective community recovery requires empowering individuals and communities to lead their own recovery journey, which increases positive outcomes but requires careful balance to avoid overwhelming or disempowering those impacted.
4. Practitioners are encouraged to recognize and accept that it is appropriate for communities to be making decisions about their futures, and this requires a shift in practitioner's mindsets.
5. Adopting an "art of hosting" approach is recommended. It requires four key practices: being truly present (pre-sensing), participating skillfully in conversations, hosting conversations effectively, and co-creating with community members.
6. Practitioners need "chaordic confidence" to persist in supporting communities in support of enabling them to take ownership of decisions that affect their lives and communities.
7. When facilitating community-led decision-making, practitioners should avoid imposing unrealistic timeframes, making assumptions about participation, tokenistic engagement, and hidden agendas.
8. Practitioners are encouraged to consider giving communities time and space, understanding community values, working with local leaders, encouraging inclusion of marginalized groups, being transparent about constraints, and fostering long-term community-led recovery.

REFERENCES

Corrigan, C. (2020). *Reflections on the Four Fold Practice of the Art of Hosting*.

Houmann, A. (2022). Exploring uncertainty by groaning in the groan zone: A pattern for emergent conversation and learning. In *Uncertainty: A catalyst for creativity, learning and development* (pp. 295–317). Springer International Publishing.

Institute of Medicine (2015). *Communities after disaster: Strategies, opportunities, and planning for recovery*. National Academy of Sciences. https://nap.nationalacademies.org/read/18996/chapter/1#ii

McKergow, M. (2015). Develop authentic leadership – Be a good host approaching leadership in a new way using the familiar techniques of hosting. *Strategic HR Review, 14*(3), 85–88.

Mitchell, L. (2019). *Considerations for governments supporting community-led recovery.* Social Recovery Reference Group, Melbourne. https://knowledge.aidr.org.au/media/7349/considerations-for-governments-supporting-community-led-recovery-final.pdf

Owen, C., Hatton, T., Mitchell, J., & Parsons, D. (2024). From hero to host: Moving beyond gendered stereotypes in emergency management leadership. *The Australian Journal of Emergency Management*, *39*(3), 66–72.

Petris, S. (2020). *Community-led recovery from extreme events sounds great – But what does it really mean?* Nous Group. https://nousgroup.com/insights/community-led-recovery/

Wheatley, M., & Frieze, D. (2011). Leadership in the age of complexity: From hero to host. *Resurgence Magazine*, 264, 4–17. www.margaretwheatley.com/articles/Leadership-in-Age-of-Complexity.pdf

18 Social License and Political Layers

INTRODUCTION

Establishing and maintaining a social license to operate is essential especially under VUCA conditions. Navigating the political landscape of competing tensions and priorities is also important for both effectiveness and survival. Therefore, understanding elements of social license and political astuteness is important for decision-makers operating under pressure.

Organizations require some degree of social license to operate, which refers to the community's acceptance of an organization's activities. According to the Ethics Centre (2018), while gaining a social license can be challenging, losing it can happen swiftly. Gaining and maintaining social license comprises three essential components: legitimacy, credibility, and trust.

COMPONENTS OF SOCIAL LICENSE

1. **Legitimacy:** This element reflects how well an individual or organization aligns with the community's norms, including legal, social, cultural, and informal regulations. An organization that adheres to these expectations is more likely to gain legitimacy.
2. **Credibility:** This refers to an organization's ability to provide accurate and transparent information to the community while honoring its commitments. A credible organization builds its reputation based on reliability.
3. **Trust:** Trust involves a willingness to be vulnerable to the actions of another, representing a strong quality of relationship that takes time and effort to cultivate. Trust is fundamental in establishing lasting connections with stakeholders.

IMPORTANCE OF SOCIAL LICENSE

Having social license to operate signifies that an organization has ongoing approval from the community, stakeholders, and political entities, ensuring public acceptance of its operations.

At an organizational level, legitimacy is often enshrined in, for example, legislation that gives the organization authority to act in certain ways. However, as recent public events have shown, not all stakeholders or community members believe an organization is legitimate just because it has been established through legislation.

To effectively act, leaders and their teams must establish legitimacy for the actions proposed, which involves gaining support from stakeholders and navigating

DOI: 10.1201/9781003566007-21

the authorizing environment. This legitimacy is crucial for securing acceptance of planned actions.

Building on the importance of legitimacy is the second element, which is that the organization, and its personnel have credibility. Credibility combines both technical and social aspects, emphasizing the need for reliable performance and responsible behavior. This means that stakeholders accept that they have the means to act in particular ways in alignment with their purposes.

The third element, trust, comes from both the legitimacy, and credibility of performance that is carried out in ways that are consistent and transparent. However, trust can only be established through shared experiences and consistent engagement. If stakeholders feel that their interests are overlooked or not prioritized, acceptance can quickly dissipate.

IMPLICATIONS FOR DECISION-MAKERS UNDER PRESSURE

While leaders and their teams may possess the necessary authorizing power to operate, their success also relies heavily on demonstrating a genuine commitment to stakeholder interests.

Building Legitimacy and Credibility

For leaders and their teams to take effective action, they must establish legitimacy within their operating environment. This legitimacy hinges on gaining stakeholder support, which allows for the acceptance of planned actions. Credibility is about meeting expectations in a way that reinforces the collective belief that the practitioners and their organizations have both the competence to be able to do what is needed and can be depended upon. Essentially, credibility involves:

- **Providing Reliable Information:** Consistently delivering accurate data and insights.
- **Honoring Commitments:** Following through on promises to reinforce stakeholder confidence.
- **Acting Responsibly:** Demonstrating ethical behavior and accountability in all actions.

The Importance of Trust

Leaders need to instill confidence among stakeholders, and they can do this by being transparent and acting in ways that are consistent with stated goals. This requires anticipating the concerns and questions that stakeholders may have and being politically astute.

UNDERSTANDING THE POLITICAL LANDSCAPE

Another critical aspect of maintaining a social license involves an awareness of the political landscape and demonstrating political astuteness. Practitioners must navigate the complexities of relationships and power dynamics within their environments.

By understanding the motivations and influences of various stakeholders, practitioners can more effectively advocate for their initiatives while ensuring that stakeholder interests align with organizational goals.

In summary, managing crises requires a multifaceted approach grounded in credibility, support, transparency, and a keen understanding of the political context. Practitioners ignore the politics of an event at their peril.

The Role of Politics in a Crisis

Aristotle's observation that "man (sic) is a political animal" highlights the inevitability of politics in all areas of life, including crisis situations. Politics and governance are deeply intertwined, and the dynamics of power often shape how crises are managed. In short, in times of crisis, political considerations become paramount.

Joel De Luca, in his book *Political Savvy* (1999), emphasized the importance of being politically astute, especially during turbulent times. He warned that many promising ideas and managerial strategies can falter due to dysfunctional organizational politics. Therefore, practitioners need to cultivate political savvy to build coalitions and address competing interests effectively (see also Chapter 11).

Political Nature of Crises

Michael Selves, who held the role of Director of Emergency Management and Homeland Security at Johnson County, Kansas, identified three reasons why emergencies possess political dimensions (Selves, 2015):

1. **Impact on People:** Emergencies have direct effects on individuals and communities, which inevitably bring political considerations into play.
2. **Public Policy Involvement:** Crises often necessitate public policy decisions, making political involvement crucial for effective management and resolution.
3. **Public and Media Interest:** Emergencies attract significant media attention, influencing public perception and political narratives.

Factors Influencing the Political Nature of Disasters

Selves (2015) also outlined several key factors that can amplify the political dimensions of specific events:

1. **Cause:** The nature of the crisis itself is a significant factor. High-impact events tend to be more politicized compared to slower-developing situations that do not initially command public attention. When a crisis has a potential human cause, such as negligence or mismanagement, it tends to provoke greater political scrutiny and media frenzy compared to natural events, which are typically not assigned blame (see also Chapter 13).
2. **Scope:** The political implications of a crisis increase with its scope. Crises requiring coordination among local, state, and federal levels of government are more likely to escalate political tensions and disputes.

3. **Public Perceptions:** Public perception can be affected by various factors, including the adequacy and timeliness of the response. Crises involving difficult decisions – such as evacuations, property seizures, or curfews – can irritate the public and heighten political implications. Moreover, if a crisis could have been mitigated or prevented through appropriate actions (e.g., flood management or disaster preparedness), it raises policy questions that are inherently political.
4. **Response Actions:** The manner in which response efforts are executed is critical. Questions around whether response actions were adequate and timely, or if mistakes were made during the response (such as delays or poor coordination), can significantly impact public satisfaction. High levels of perceived dissatisfaction can amplify the political nature of the situation, leading to greater scrutiny from both the public and the media.

Crises Have Political Consequences

As noted by Selves (2015), crises possess an inherent capacity to influence the political landscape of communities, states, or nations. Such events can shift public perceptions regarding the competence and empathy of political leaders, making the public increasingly sensitive to any criticisms related to response and relief efforts.

Such things as whether the political players are on the same "team", whether previous partisan divisions existed or whether policy disagreements have been a factor in the past all affect, to one degree or another the political climate. Are the political players adversaries generally and disinclined to cooperate or do they share a common political agenda? Where are we in the election cycle?

One of the primary challenges for decision-makers during a crisis is the need to balance necessary actions with what will be acceptable with their constituencies. While effective crisis management is generally well-received, certain necessary measures – such as imposing curfews, limiting the sale of certain products, mandating evacuations, closing businesses, or enforcing quarantines – can be deeply unpopular. Practitioners must navigate these difficult decisions with care, ensuring that the essential actions taken to protect public safety do not erode public trust.

The involvement of political staffers and policy advisors in crisis situations can further complicate the political dynamics at play. These personnel are typically not engaged in emergency management and become involved only when their elected officials turn their attention to a crisis. As such, maintaining open lines of communication and providing valuable insights to political staff can enhance the overall effectiveness of response efforts.

Politics Can Have Disastrous Consequences

The impact of political considerations on managing a crisis is often profound, sometimes resulting in negative consequences. These can range from minor inconveniences, such as tours by politicians at disaster sites, to significant obstructions in achieving response objectives, such as lack of funding or refusal to grant necessary authority. There are rare but serious cases where political motivations can lead to unethical actions, such as misusing emergency powers for personal gain.

It is vital for practitioners to recognize and proactively manage the potential for political considerations to complicate management of high-consequence VUCA-related events. Implementing well-defined operating plans and procedures can significantly mitigate the risks of politicization during response and recovery efforts.

Politicians, Like Nature, Abhor a Vacuum

For practitioners, understanding the political dimensions of VUCA situations is essential. While excessive political interference can be problematic in some instances, most practitioners find that if operations are stable and coordinated, political stakeholders tend to respect established protocols. It is during times of perceived chaos or disorganization that political leaders are more likely to step in and potentially disrupt the response efforts.

Selves (2015) suggests that practitioners possess the power to shape their interactions with the political sphere by maintaining professionalism and understanding crisis management processes. Effective communication of these principles – before, during, and after a crisis – is essential. Key strategies include:

- **Developing Comprehensive Plans:** Crisis management plans should be robust, adaptable, and ready to address any contingencies that may arise during an event.
- **Building Relationships:** Establishing strong relationships with political stakeholders and the media ahead of an emergency is crucial. Political actions are often reactions to public perceptions, heavily influenced by media narratives.
- **Professional Conduct:** Conducting operations with professionalism fosters confidence among stakeholders, allowing emergency managers to maintain control and authority during crises.

In summary, practitioners can create a framework for coordination and understanding that minimizes the potential for negative political interference. It is incumbent upon decision-makers to be proactive in establishing strong relationships and fostering trust, as this can significantly impact their effectiveness during crises. Ignoring the political dimension of crisis can jeopardize not only their authority but also the well-being of the communities they serve.

Understanding the distinction between power and influence is also vital for successful leadership. Leaders often wield power subtly; for example, positions of authority, such as wearing a uniform or holding a title like Mayor, command a level of respect and deference. As discussed in Chapter 9, leaders and followers can succumb to silence under uncertain conditions. By understanding the political dimensions of their actions and decisions, practitioners can better address the needs of their communities while fostering trust and maintaining their social license to operate.

STRATEGIES FOR PRACTITIONERS: POLITICAL ASTUTENESS FOR LEADERSHIP IN VUCA

In high-consequence VUCA-related events, practitioners navigate a challenging landscape characterized by a mix of power, influence, and politics, all framed by legislation, policy, public expectation, and various hidden agendas. To be effective, practitioners must operate within this context while maintaining a strong ethical stance and compliance with relevant guidelines (see also Chapter 6). In short, they need political astuteness.

The Five Dimensions of Political Astuteness

According to Hartley et al. (2013), political astuteness is considered a "meta-competency" comprising five key dimensions:

1. **Personal Skills:** Self-awareness of one's motives and behaviors, combined with the ability to exercise self-discipline.
2. **Interpersonal Skills:** The capacity to engage with individuals across various levels of authority, influencing them through trust and relationship-building.
3. **Situational Awareness:** The ability to read people and situations, recognizing underlying dynamics and agendas, including understanding when one's own role may be perceived as a threat.
4. **Network Building:** The capability to form alliances and collaborate despite conflicting interests, effectively leveraging diversity to drive constructive action.
5. **Strategic Scanning:** The ability to focus on long-term issues and challenges, rather than becoming overwhelmed by immediate pressures.

McConnell's Perspectives on Success in Disaster Management

Allan McConnell (2011) emphasizes that perceptions of success in handling crises are subjective and vary among stakeholders. Different interpretations of how effectively an event was managed can coexist based on the interests and values of those involved. According to McConnell, perceptions of crisis management can be categorized into several storylines:

- **Outright Success:** The management of the event was commendable under extraordinary circumstances, garnering awards or recognition.
- **Outright Failure:** The crisis was poorly managed, and those responsible should face consequences.
- **Durable Success:** The response achieved more positive outcomes than negative, despite imperfections.
- **Conflicted Success:** Many aspects were handled well, but significant missteps were also apparent.

Tool 18.1 assists practitioners to evaluate the role of politics in a VUCA-related event and the ways in which their efforts might be perceived.

TOOL 18.1 ANALYZING THE POLITICS OF A CRISIS EVENT

Consider a crisis you have been involved in and jot down some notes addressing the following:

1. **How did the political aspects play out?** What dynamics emerged among stakeholders, and how did these influence the management of the crisis?
2. **How was the response perceived using McConnell's frames?** In which category do you believe it fits, and why?
3. **Why was this the case?** Reflect on stakeholder interests, external perceptions, and events that may have shaped these views.

By analyzing these factors, practitioners can gain deeper insights into the interplay of politics and crisis management, enhancing understanding and preparedness for future crises. Reflecting on recent crises encountered or observed can help to analyze the political dimensions at play.

Consider the following questions:

1. **How did the disaster unfold?** Did it escalate rapidly, or were there observable warning signs?
2. **What dimensions were involved?** Consider aspects such as issues of blame, indirect effects on connected parties, and opportunistic claims for change.
3. **What were the major political concerns in the sector?** Examining the emergency through the lens of the following considerations can yield valuable insights:
   - **Reputation:** How did the events affect the public image of involved organizations?
   - **Resources:** Were there sufficient resources allocated, or did shortages become apparent?
   - **Champions and Advocates:** Who supported effective response efforts?
   - **Public Opinion and Expectation:** How did community sentiment shape the response?
   - **Responsibility:** Who was held accountable for the crisis outcomes?
   - **Economics:** What economic considerations influenced decisions?
   - **Future Emergency Events:** How did this event prepare stakeholders for possible future crises?

MAIN POINTS

- The interplay of political dynamics, stakeholder interests, and public perception significantly influences crisis management outcomes.
- Establishing trust and transparency is crucial for effective crisis management to maintain their social license to operate.
- Responses to crises can be evaluated through different lenses, such as McConnell's (2011) frames of success, illustrating the subjective and political nature of stakeholder evaluations.
- Prioritizing stakeholder interests is essential for fostering credibility during crises.
- The ability to navigate these complexities is key to the effectiveness of future responses and organizational resilience in emergencies.

REFERENCES

DeLuca, J. R. (1999). *Political savvy: Systematic approaches to leadership behind-the-scenes*. Evergreen Business Group.

Ethics Centre (2018). *Ethics explainer: Social license to operate*. Retrieved March 24, 2025, from https://ethics.org.au/?s=social+licence

Hartley, J., Alford, J., Hughes, O., & Yates, S. (2013). *Leading with political astuteness: A study of public managers in Australia, New Zealand and the United Kingdom*.

McConnell, A. (2011). Success? Failure? Something in-between? A framework for evaluating crisis management. *Policy and Society*, *30*(2), 63–76.

Selves, M. (2015). *The politics of disaster: Principles for local emergency managers and elected officials*. Retrieved January 18, 2023, from https://www.scribd.com/document/383778741/The-Politics-of-Disaster-1

Conclusion

INTRODUCTION

In this book, we have argued that decision-makers operating in volatile, uncertain, complex, and ambiguous (VUCA) environments need to have complex sense-making skills to appropriately interpret and manage current and future challenges. According to Elsner et al. (2025), who captured insights from over 900 experts worldwide, we are living in one of the most divided times since the Cold War. Growing divisions and rising fragmentation are transforming international relations, raising doubts about whether current governance and multilateral agreements can effectively address the shared challenges we face. In various spheres – such as geopolitics, humanitarian concerns, economic ties, and environmental, social, and technological issues – global collaboration is declining, and this trajectory is anticipated to continue in the years ahead (Elsner et al., 2025).

Weather events are expected to become more extreme, and the human impact on the landscape risks enabling more biodiversity loss and, in some cases, system collapse. At the same time, the increasing use of digital platforms concentrates interdependencies leading to greater vulnerabilities in critical infrastructure such as electricity and water supply.

While the use of artificial intelligence (AI), for example, offers significant prospective benefits, it simultaneously increases security risk and expands the threat landscape across both physical and digital domains (Elsner et al., 2025). In addition, the Federation of European Risk Management Associations (FERMA, 2025), a peak body for 23 national risk management associations in 22 European countries, noted that the increasing use of AI in critical sectors such as the finance system, healthcare and energy sectors creates new potential points of failure. These changes are not typical "black swan" risks, but indeed are "grey rhinos" (FEMA, 2024, p. 7). That is "events that are high impact and highly probable, but which are often ignored or overlooked until it is too late".

We hope that Chapter 1, which discussed the nature of volatility, uncertainty, complexity, and ambiguity provides a useful context for practitioners. Volatility is an unstable or unpredictable situation that involves frequent change. Uncertainty is a situation characterized by a lack of knowledge, not as to cause and effect, but pertaining to whether an event is significant enough to be a meaningful cause. There is a lack of adequate information to predict what may happen, and decision-makers need to be flexible and expand data collection and processing resources. Complexity and uncertainty contribute to a range of dilemmas and pressures for decision-makers and teams. A key challenge for practitioners is how to identify and articulate elements of VUCA in their situations, which is where Tool 1.1 *Identifying volatility in the decision-making context* may be useful.

THE SELF AND THE CHANGING NATURE OF WORK AND WORK ORGANIZATION

One of the elements that will be critical in the future is how organizations maintain and develop their people. A report by the US Federal Emergency Management Agency (FEMA, 2024) noted that in the emergency services sector, for example, retirement is outpacing recruitment in many areas, and the sector faces burnout and stress partly due to the growing demands of increased intensity, and year-round pace of engagement leading to strains in worker stress. Across the globe, there has been a greater emphasis on the professionalization of the emergency services sector (e.g., Dippy, 2025; Kapucu et al., 2024), and this may attract more people into the emergency and disaster management sector. However, as we have argued, the responsibilities of managing in VUCA situations are now far wider than just emergency response. It is likely, therefore, that more personnel in many organizations will face the same stressors and risks of burnout because of the changes occurring at local regional and national levels.

We hope that the chapters included in this book on self-awareness, including on understanding the impacts of stress (see Chapters 2–4) and what can be done to increase resilience, will assist practitioners facing those challenges. As Chapter 2 discussed, effective decision-making in (VUCA) contexts requires a deep understanding of the interplay between pressure, stress, and cognitive processes. Chapter 3 outlined the impacts of stress on decision-making and proposed that adopting strategies such as mindfulness, deliberate recovery practices, and self-care can assist practitioners to optimize their performance and that of their teams. Chapter 4 highlighted the intricate relationship between the mind and body in decision-making, particularly under pressure. We hope that the tools provided to support *identifying sources of pressure at work* (Tool 2.1); *strategies to practice counting down to calm* (Tool 3.1); and *developing mindful body awareness* (Tool 4.1) are helpful to practitioners.

FERMA (2025) also notes that organizations will be required to pay increasing attention to soft skills, such as leadership development and emotional intelligence (EI). In a world increasingly dominated by VUCA contexts and where AI is anticipated to replace routine tasks, the role of people in managing complexity and unexpected developments will become more important. Chapter 5 discussed the pivotal role EI plays in enhancing decision-making quality, particularly under high-stress conditions. By recognizing and regulating one's emotions, as well as being aware of potential "red flag conditions" practitioners can make more informed decisions. Cultivating high EI is, therefore, an asset for decision-makers operating under pressure, leading to improved outcomes and enhanced team performance. Tools 5.1 and 5.2 provide strategies for practitioners to enhance their *emotional awareness* and to *practice self-compassion* in challenging conditions.

Technological advancement will not just require upskilling but will also transform the nature of work in many sectors. The incorporation of AI in emergency management, for example, has opened new avenues for improving response and recovery efforts during crises. By utilizing AI technology, emergency management agencies can enhance decision-making, optimize resource distribution, and elevate overall disaster response effectiveness. However, as AI's impact in this field grows, it is vital to acknowledge and tackle the ethical issues and challenges that come with its implementation. We hope Chapter 6 including ethical considerations in the deployment of AI systems provides some useful insights and strategies for practitioners. Tool 6.1 supports practitioners in being able to *identify red flag conditions,* and Tool 6.1 provides *guidance to support ethical decision-making.*

Effective leadership in VUCA situations hinges on embracing paradoxical tensions and cultivating key leadership qualities. We hope that Chapter 7, which explored challenges such as overbearing leadership, blind spots, biases, and micropolitics, enables practitioners to adapt their strategies to promote inclusive engagement, deliberative decision-making, and empowered team contributions. Tool 7.1 encourages self-reflection on *three mind traps* leaders can avoid when operating in VUCA situations.

THE SOCIAL CONTEXT: TEAM EFFECTIVENESS AND STAKEHOLDER ENGAGEMENT

In VUCA situations, team situation awareness becomes critical. Chapters 8–10 highlighted strategies to cultivate high-performing teams, which are essential when operating in uncertainty. Adopting practices outlined by Hackman in Chapter 8, such as setting a compelling direction, creating enabling teamwork structures, and launching a team for success, can significantly enhance team effectiveness. Using Tool 8.1 practitioners are encouraged to *assess the enabling conditions* of their own teams. Changing the ways in which questions about uncertainty are asked (Tool 8.2) helps to encourage voice for team members.

Building psychological safety (Chapter 9) is an important precursor to team effectiveness. Tool 9.1 provides a resource for practitioners to *assess psychological safety* within their own teams, and Tool 9.2 provides a self-assessment tool practitioners can use to *assess the quality of team talk for psychological safety.* It is also important to recognize that not all teams are immediately effective and are sometimes dysfunctional. Chapter 10 explores dysfunction and provides suggestions for practitioners to diagnose impediments to team effectiveness. Tool 10.1 provides support in *negotiating conflict creatively and using compassionate accountability.*

Chapters 11 and 12 took up the importance of strategic communication and stakeholder engagement in decision-making and leadership, particularly under pressure. Key actions were proposed to address issues collaboratively, highlighting the benefits of resolving conflicts and fostering cohesion among diverse stakeholders. It is important to recognize that each stakeholder operates from a unique frame of reference, or mental model and that effective leadership requires understanding diverse perspectives. Tools to support creating *narratives with influence* (Tool 11.1) and strategies to *map stakeholder influence to identify potential alliances* (Tool 12.1) were provided.

THE SITUATION: UNCERTAIN FUTURES AND RISK

Recent events have underscored the speed at which situations can escalate leading to new risks, However, it is ironic that, as situations are increasing in their volatility and complexity, decision-makers frequently succumb to short-term thinking and cognitive biases, which obstruct their capability to foresee and address emerging risks. According to FRMA (2024), this can result in a form of "strategic blindsight". That is, in an era marked by a pervasive awareness of potential existential threats, this feeling of uncertainty may ironically intensify avoid thinking too far ahead, and thus to an emphasis on the short to medium term.

Chapter 13 explored what can happen when organizations face existential threats to their survival and provided a typology of the kinds of strategies organizational leaders can apply, depending on whether the crisis was one that was internal or externally generated and whether it was based on intentional or unintentional acts. Under these circumstances, organizational leaders require robust governance practices that involves managing internal and external stakeholder expectations. Tool 13.1 provided a means of keeping *a strategic focus.*

The FERMA report noted that it is vital that practitioners can draw on the skills needed for foresight, including scenario planning to navigate through dynamic events. We hope that the chapters included in this book on navigating uncertainty (Chapter 14), anticipating futures thinking (Chapter 15), and engaging in scenario planning will yield helpful insights.

Given the expectation that future work will be transformed by convergent technologies and the need to overcome the challenges of burnout identified by FEMA (2024), it can be anticipated that managing VUCA situations resulting in disasters and crises will not just be the province of emergency service personnel but also citizens and communities. Empirical evidence consistently demonstrates that collaborative efforts between citizens and local emergency management professionals significantly reduce the negative impacts of disasters and other crises (Choi, 2025). It is important, therefore, for practitioners to be able to better understand the potential impacts of psycho-social trauma of communities that have experienced a crisis or a disaster, which is where Tool 16.1 *assessing the potentials for social de-bonding* can be a useful resource (see Chapter 16).

Daddoust et al. (2021) also highlighted the opportunities, challenges and risks of spontaneous volunteers who turn up unannounced to help in disasters. Volunteers are a significant and important human resource component of the timely emergency workforce; their assistance is often readily available and without direct costs. Many disasters would have been far more lethal and consequential without their engagement. We hope that Chapter 17,which explored the intersection of emergency services personnel and citizens impacted by a crisis, may be helpful to those practitioners hoping to host conversations and community-led decision-making. Tool 17.1 here provides *considerations when facilitating community-led decision-making.*

Elsner et al. (2025) noted that the expansion of conflicts, a variety of extreme weather events intensified by climate change, extensive societal and political polarization, and escalating technological advancements are exacerbated by the spread of false or misleading information. These changes can sow doubt among the general public worldwide about what is happening in VUCA situations. The growing volume of AI-generated content also makes divisive misinformation and disinformation more pervasive. All these issues raise the importance of how practitioners create appropriate narratives for sense-making (see Chapter 11), navigate uncertainty (Chapter 14), and manage and maintain social license. Chapter 18 included a tool

(Tool 18.1) to support practitioners in their *analysis of the politics in VUCA events* to support political astuteness under these conditions.

There is no doubt that we live in challenging times and that all indicators point to a continuation and escalation of VUCA in our lives and which provide evolving threats and opportunities. As interdependencies and digital platforms become more tightly coupled those threats and opportunities are characterized by systemic risk (FERMA, 2025). Systemic risk is when a crisis causes the breakdown of an entire system not just one of its constituent parts (FERMA, 2025). These can take the form of, for example, a cyber war that could cripple financial markets within hours; a water conflict that could generate a world-wide food crisis; technological blocks that could incapacitate global trade; or trigger disruption to energy supplies (FERMA, 2025).

This sourcebook provides valuable support for practitioners in uncertain times. It provides insights into a range of factors that impact the performance of decision-makers when confronted by high uncertainty. For astute leaders, the sourcebook contains many tools and techniques that act like keys unlocking the doors to opportunity, giving practitioners the edge for success in a world of uncertainty and challenge.

REFERENCES

Choi, J. (2025). Collective coproduction of emergency services: Exploring key determinants using machine learning. *International Public Management Journal*, *28*(5), 626–649. https://doi.org/10.1080/10967494.2025.2456863

Daddoust, L., Asgary, A., McBey, K. J., Elliott, S., & Normand, A. (2021). Spontaneous volunteer coordination during disasters and emergencies: Opportunities, challenges, and risks. *International Journal of Disaster Risk Reduction*, *65*, 102546.

Dippy, R. (2025). Professionalism: Application of the concept of disciplinarity to emergency management. *The Australian Journal of Emergency Management*, *40*(3), 6–16.

Elsner, M., Atkinson, G., & Zahidi, S. (2025). *Global risks report, 2025.* World Economic Forum. Retrieved July 25, 2025, from https://www.weforum.org/publications/global-risks-report-2025/

FEMA (2024, November). *Strategic foresight 2050, final report.* Retrieved June 2, 2025, from https://www.fema.gov/sites/default/files/documents/fema_strategic-foresight-2050-final-report.pdf

FERMA (2025). *NEXT: New exposure trends report, 2025.* Federation of European Risk Management Associations: Anticipating Changes Shaping the Future. Retrieved August 2, 2025, from https://ferma.eu/next-report-2025/

Kapucu, N., Parkin, A., Lumb, M., & Dippy, R. (2024). Crisis coordination in complex intergovernmental systems: The case of Australia. *Public Administration Review*, *84*(3), 389–399.

Visave, J. (2024). AI in emergency management: Ethical considerations and challenges. *Journal of Emergency Management and Disaster Communications*, *5*(1), 165–183.

Index

For Product Safety Concerns and Information please contact our EU representative GPSR@taylorandfrancis.com Taylor & Francis Verlag GmbH, Kaufingerstraße 24, 80331 München, Germany